Amazon Mission

Building Math

Integrating Algebra & Engineering

The *Building Math* project was funded by a grant from the GE Foundation, a
philanthropic organization of the General Electric Company that works to strengthen
educational access, equity, and quality for disadvantaged youth globally.

The interpretations and conclusions contained in this book are those of the
authors and do not represent the views of the GE Foundation.

1   2   3   4   5   6   7   8   9   10
ISBN 978-0-8251-6878-9
J. Weston Walch, Publisher
40 Walch Drive
Portland, ME 04103
www.walch.com
Printed in the United States of America

TABLE OF CONTENTS

Introduction . . . . *v*

Organization and Structure of *Amazon Mission* . . . . *vii*

Building Math: Pedagogical Approach, Goals, and Methods . . . . *viii*

Enduring Understandings . . . . *ix*

Amazon Mission Overview: Story Line and Learning Objectives . . . . *x*

Assessment Opportunities and Materials Lists . . . . *xi*

Amazon Mission Master Materials List . . . . *xiv*

Common Core and ITEEA Standards Correlations . . . . *xv*

Pacing Planning Guide . . . . *xxi*

"The Law for the Wolves" Team-Building Activity 1

Amazon Mission Prerequisite Math Skills. 3

Writing Heuristics, or Rules of Thumb 21

Amazon Mission Introduction 22

Introducing the Engineering Design Process (EDP) 26

Design Challenge 1: Malaria Meltdown! 32

Design Challenge 2: Mercury Rising! 73

Design Challenge 3: Outbreak! 108

Resources & Appendices 147

   EDP: Engineering Design Process 148

   Math and Engineering Concepts 150

   Important Vocabulary Terms 150

   Rubrics 152

   Student Work Samples 160

   Appendix: Net of a Rectangular Prism 164

Answer Key 165

INTRODUCTION

Welcome to the second edition of *Building Math*. This revised program incorporates feedback from instructors and is correlated to the Common Core State Standards (CCSS).

Building Math is a unique program that integrates real-world math and engineering design concepts with adventurous scenarios that draw students in. The teacher-tested, research-based activities in this program enforce critical thinking skills, teamwork, and problem solving, while bringing students' classroom experiences in line with the Common Core.

Each set of three activities (Design Challenges) forms one unit. The unit's activities are embedded within an engaging fictional situation, providing meaningful contexts for students as they use the engineering design process and mathematical investigations to solve problems. There are three units, and each unit takes about three weeks of class time to implement.

WHAT'S INCLUDED IN THE BUILDING MATH PROGRAM

The instructional materials include reproducible student pages, teacher pages, a DVD of classroom videos for teacher professional development and a Java applet used as a computer model in one of the activities, and a poster showing the engineering design process. The full program is also provided as an interactive PDF on an accompanying CD. This includes all of the content from the book as well as expanded CCSS correlations. The CD is intended to facilitate projecting materials in the classroom and/or printing student pages. You can also print out transparencies from the instructional pages if it suits your needs.

WHY IS ENGINEERING EDUCATION IMPORTANT?

The United States is faced with the challenge of increasing the workforce in quantitative fields (e.g., engineering, science, technology, and math). Schools and teachers play a pivotal role in this challenge. Currently, many students (mostly from underrepresented groups) are not graduating from high school with the necessary math skills to continue studies in college in these quantitative fields. Colleges and industries such as Tufts University and General Electric realize that more active participation in the pre-K–12 system is needed, and have put together this innovative program to help teachers increase math and engineering content in the middle-school curriculum.

(continued)

Algebra and engineering are critical fields that are worth combining. Algebraic reasoning acts as a foundation for higher levels of math learning in secondary and tertiary education, and introducing students to engineering is a way to show them how math is used as a discipline of study and a career path.

BACKGROUND

Building Math was a three-year project funded through the GE Foundation in partnership with the Museum of Science in Boston. One goal of the project was to provide professional development for middle-school teachers in math and engineering, and to explore alternative teaching methods aimed at improving eighth grade students' achievement in algebra and technology. Another project goal was to develop standards-based activities that integrate algebra and engineering using a hands-on, problem-solving, and cooperative-learning approach.

The resulting design challenges were tested by teachers in ten Massachusetts schools that varied in type (public, charter, and independent), location (urban, suburban, rural), and student demographics. Subsequently, hundreds of teachers all over the country used the materials with their students. Their experiences have informed this second edition of *Building Math,* which includes correlations to Common Core State Standards. Further, this edition includes specific, optional suggestions to allow teachers to address additional aspects of the CCSS.

PARTICIPATING RESEARCHERS AND SCHOOLS

Project Investigators: Dr. Peter Y. Wong and Dr. Bárbara M. Brizuela
Project Coordinators: Lori A. Weiss and Wendy Huang
Pilot Schools and Teachers:
- Ferryway School (Malden, MA): Suzanne Collins, Julie Jones
- East Somerville Community School (Somerville, MA): Jack O'Keefe, Mary McClellan, Barbara Vozella
- West Somerville Neighborhood School (Somerville, MA): Colleen Murphy
- Breed Middle School (Lynn, MA): Maurice Twomey, Kathleen White
- Fay School (Southborough, MA): Christopher Hartmann
- The Carroll School (Lincoln, MA): Todd Bearson
- Knox Trail Junior High School (Spencer, MA): Gayle Roach
- Community Charter School of Cambridge (Cambridge, MA): Frances Tee
- Mystic Valley Regional Charter School (Malden, MA): Joseph McMullin
- Pierce Middle School (Milton, MA): Nancy Mikels

ORGANIZATION AND STRUCTURE OF *AMAZON MISSION*

Throughout this reproducible book, you will find teacher guide pages followed by one or more student activity pages. Each page is labeled as either a student page or a teacher page.

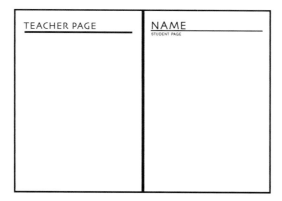

Answers to all activities and discussion questions are found in the answer key at the back of the book.

TIPS, EXTENSIONS, OPTIONAL CCSS ENHANCEMENTS, and **ASSESSMENTS** are labeled in gray boxes.

 INTERESTING INFO is provided in white boxes. This provides additional related information and resources that you may want to share with students.

The following labels are used to indicate whether you will be addressing the whole class, teams, individuals, or pairs:

CLASS **INDIVIDUALS** **TEAMS** **PAIRS**

THE ENGINEERING DESIGN PROCESS

Activities align with the eight-step engineering design process. See page 26 for a lesson plan to introduce the steps.

TABLES AND GRAPHS

Tables and graphs are numbered according to their order of appearance in each design challenge. Those beginning with 1 correspond to Design Challenge 1. Those beginning with 2 correspond to Design Challenge 2. Those beginning with 3 correspond to Design Challenge 3.

BUILDING MATH: PEDAGOGICAL APPROACH, GOALS, AND METHODS

The approach of the *Building Math* program is to engage students in active learning through hands-on, team-based engineering projects that make learning math meaningful to the students (Goldman & Knudson).

The goal of *Building Math* is to encourage the development of conceptual, critical, and creative-thinking processes, as well as social skills, including cooperation, sharing, and negotiation by exhibiting four distinctive methods (Johnson, 1997).

1. **Contextual learning**—Each *Building Math* book involves students in a story line based on real-life situations that pose fictional, but authentic, design challenges. Design contexts invite students to bring ideas, practices, and knowledge from their everyday lives to classroom work. Students apply math skills and knowledge in meaningful ways by using math analysis to connect inquiry-based investigations with creating design solutions.

2. **Peer-based learning**—Students work together throughout the design process. The key to peer-based learning is high amounts of productive, on-task verbalization. By verbalizing their thoughts, students listen to their own and others' thinking. This allows them to evaluate and modify one another's thinking and defend their own ideas. Verbalization also contributes to more precise thinking, especially when teachers use effective questioning techniques to ask students to explain and analyze their and others' reasoning.

3. **Activity-based practice**—*Building Math* uses design challenges (design and construction of a product or process that solves a problem) to focus peer-based learning. Students conduct experiments and systematic investigations; use measuring instruments; carefully observe results; gather, summarize, and display data; build physical models; and analyze costs and trade-offs (Richards, 2005).

4. **Reflective practice**—*Building Math* activities include questions, rubrics, and self-assessment checklists for students to document and reflect on their work throughout each design stage. Teams summarize and present their design solutions to the class, and receive and offer feedback on others' solutions.

REFERENCES

Goldman, S., & Knudsen, J. Learning sciences research and wide-spread school change: Issues from the field. Paper presented at the International Conference of the Learning Sciences (ICLS).

Johnson, S. D. 1997. Learning technological concepts and developing intellectual skills. International Journal of Technology and Design Education, 7, 161–180.

Richards, L. G. 2005. Getting them early: Teaching engineering design in middle-schools. Paper presented at the National Collegiate Inventors & Innovators Alliance (NCIIA). http://www.nciia.org/conf05/cd/supplemental/richards1.pdf

ENDURING UNDERSTANDINGS

In their book, *Understanding by Design,* Wiggins and McTighe advocate that curricula can be built by identifying enduring understandings. An enduring understanding is a big idea that resides at the heart of a discipline, has lasting value outside of the classroom, and requires uncovering of abstract or often misunderstood ideas. The list below details the enduring understandings addressed in *Building Math*. Teachers can consider these to be the ultimate learning goals for the *Building Math* series.

ENGINEERING AND TECHNOLOGICAL LITERACY

1. Technology consists of products, systems, and processes by which humans modify nature to solve problems and meet needs.

2. Design is a creative planning process that leads to other useful products and systems.

3. There is no perfect design.

4. Requirements for design are made up of criteria and constraints.

5. Design involves a set of steps, which can be performed in different sequences and repeated as needed.

6. Successful design solutions are often based on research, which may include systematic experimentation, a trial-and-error process, or transferring existing solutions done by others.

7. Prototypes are working models that can later be improved to become valuable products. Engineers build prototypes to experiment with different solutions for less cost and time than it would take to build full-scale products.

8. Trade-off is a decision process recognizing the need for careful compromises among competing factors.

MATH

1. Math plays a key role in creating technology solutions to meet needs.

2. Mathematical models can represent physical phenomena.

3. Patterns can be represented in different forms using tables, graphs, and symbols.

4. Graphs are useful to visually show the relationship between two variables.

5. Measurement data are approximated values due to tool imprecision and human error.

6. Repeated trials and averages can build one's confidence in measurement data.

7. Mathematical analysis can lead to conclusions to help one make design decisions to successfully meet criteria and constraints.

8. Analyzing data can reveal possible relationships between variables, and support predictions and conjectures.

REFERENCES

International Technology and Engineering Educators Association (ITEEA). 2007. *Standards for Technological Literacy: Content for the Study of Technology.* (Third ed.). Virginia.

Wiggins, G., and J. McTighe. 2005. *Understanding by Design.* (2nd ed.). Prentice Hall.

AMAZON MISSION OVERVIEW: STORY LINE AND LEARNING OBJECTIVES

	Design Challenge Overview	Students will:
DESIGN CHALLENGE 1: MALARIA MELTDOWN!	Students are responding to the needs of the Yanomami people in the Amazon. In their first challenge, they are to design a medicine carrier that can successfully transport malaria medicine. The carrier should keep the medicine within certain temperature constraints to protect it from heat, be rugged enough to prevent an egg from breaking when dropped, and be as low in cost as possible.	• calculate and interpret the slope of a line • graph a compound inequality • conduct two controlled experiments • collect experimental data in a table • produce and analyze a line graph that relates two variables • distinguish between independent and dependent variables • determine when it's appropriate to use a line graph to represent data • list combinations of up to five layers of two different kinds of materials • draw a three-dimensional object and its net • find the surface area of a three-dimensional object • apply the engineering design process to solve a problem
DESIGN CHALLENGE 2: MERCURY RISING!	As students arrive at the village, the Yanomami meet them with a new challenge—to design a water filter that can filter out at least 75% of the mercury in the freshwater near the mining operation. To do so, students research different sizes of Mercatrons, mercury-absorbing spheres. Students find which ones would meet the criteria of being low in cost and still effective at removing at least 75% of mercury from water. Students also calculate minimum and maximum flow rates for water and experiment with different factors that influence the flow rate.	• calculate the surface area of a sphere using a formula • solve a multistep problem • convert measurement units (within the same system) • use proportional reasoning • write a compound inequality statement • graph and analyze the relationship between two variables • determine when it's appropriate to use a line graph to represent data • design and conduct a controlled experiment • apply the engineering design process to solve a problem
DESIGN CHALLENGE 3: OUTBREAK!	The Yanomami are vulnerable to infectious diseases brought by outsiders. Students are challenged to select from a list of interventions to form a virus containment plan. Students conduct simulations of virus spread under different conditions, calculate percentage rate of infection with different combinations of interventions, and use their results to design a virus containment plan that would keep the percentage of infected villagers to no more than 25% for 30 days and be as low in cost as possible.	• identify and extend exponential patterns • generalize and represent a pattern using symbols • graph simulation data and describe trends • calculate compound probabilities • use a computer model • apply the engineering design process to solve a problem

ASSESSMENT OPPORTUNITIES AND MATERIALS LISTS

The tables on the next three pages list the opportunities for formative assessment by using rubrics, probing students' thinking during class time, and reviewing student responses to certain questions. The tables also show the materials needed for each design challenge. The numbers in the assessment column refer to the steps of the engineering design process.

Assessment	Materials
• **2. Research:** Assess whether students can describe the relationship represented in the graph by the variables in the *x*- and *y*-axes. • **2. Research:** Assess whether students can calculate the slope on the graph between two data points and interpret the meaning of the slope. • **2. Research:** Assess whether students can describe changes in the rate (i.e., changes in slope) and explain what these changes mean. • **2. Research:** Assess whether students can represent the temperature criterion as a compound inequality and represent that inequality as a shaded area on the graph. • **2. Research:** Assess whether students can interpret the graph to see how well the carrier performance meets the temperature criterion. • **2. Research:** Assess whether students can sketch a graph that would meet the temperature criterion. • **2. Research:** Assess whether students can make a complete graph and correctly represent the experimental data. Use the rubric on page 152. • **2. Research:** Assess whether students can apply the findings of their research to create a rule of thumb to help meet design criteria. • **4. Choose:** Assess engineering drawing based on quality and communication using the rubric on page 155. • **4. Choose:** Assess how well students can mentally unfold a 3-D shape to draw its net. • **4. Choose:** Assess whether students can come up with the logical steps required to find the total cost of the materials for their prototypes. • **5. Build:** Assess model/prototype artifact based on craftsmanship and completeness. Use the rubric on page 157. • **6–8. Test, Communicate, Redesign:** Assess written responses and student observations during test, communicate, and redesign steps based on model performance, completeness, and quality of reflection. Use the rubric on page 158. • **Individual Self-Assessment Rubric:** Students can use the checklist on page 70 to determine how well they met behavior and work expectations. • **Team Evaluation:** Students can complete the questions on page 72 to reflect on how well they worked in teams and celebrate successes, as well as make plans to improve teamwork. • **Student Participation Rubric:** Make copies of the rubric on page 159 to score each student's participation in the design challenge.	For each team: • 1 pack colored pencils • 1 digital thermometer • 1 cup (8 oz) of ice • 1 stopwatch • 2–5 pieces of each material (15 cm × 15 cm): corrugated cardboard, foam board, bubble wrap, and aluminum foil • large pieces of materials (see above) to construct carrier prototype • 2 transparencies • 1 pack transparency markers (min. 4 colors) • 1–2 pairs of scissors • 1 roll heavy-duty packing tape • 1 meterstick • 1 net of a rectangular prism (see Appendix) on standard letter-sized cardstock (optional) • egg (raw) • 1 resealable sandwich bag For the class: • 1 overhead projector • sheet of chart paper

DESIGN CHALLENGE 1: MALARIA MELTDOWN!

Assessment	Materials
• **2. Research:** Assess logical reasoning by asking students to figure out the steps they would need to find out which size spheres can meet the surface-area criterion at the lowest cost possible rather than just telling them the steps.	For each team: • 1–2 calculators • 1 pin • 1 small nail
• **2. Research:** Assess whether students can convert units to desired equivalent forms.	• 1 large nail • 1 chopstick
• **2. Research:** Assess whether students can represent the flow rate criterion in compound inequality form.	• 1 ruler (metric) • 1 measuring cup with 250-mL line
• **2. Research:** Assess whether students can make a complete graph and correctly represent the experimental data. Use the rubric on page 152.	• 1 stopwatch • 1 bottle of water (about 500 mL)
• **2. Research:** Assess whether students can describe the relationship represented in the graph by the variables in the *x*- and *y*-axes.	• 1 roll packing tape • 1–2 pairs of scissors
• **2. Research:** Assess whether students can apply the findings of their research to create a rule of thumb to help meet design criteria.	• 2 small plastic bags • 6 plastic stirrers
• **2. Research:** Students are asked to choose a factor that they think will affect the flow rate, make a prediction, test their prediction by varying the factor while keeping all other conditions the same, record the results, graph the results, and analyze the results to create a rule of thumb to share with the class. The task can be assessed based on: overall completeness, quality of test procedures, graph completeness and quality, and data analysis. Use the rubric on pages 153–154.	• 6 plastic straws • 8 small (8-oz) cups • 4 large (12-oz) foam cups • 90 plastic spheres 1.2 cm in diameter (½ or ⅜ inch); Mardi Gras beads work well
• **4. Choose:** Assess engineering drawing based on quality and communication. Use the rubric on page 156.	• water basin or pail • sheet of chart paper
• **5. Build:** Assess model/prototype artifact based on craftsmanship and completeness. Use the rubric on page 157.	
• **6–8. Test, Communicate, Redesign:** Assess written responses and student observations during test, communicate, and redesign steps based on model performance, completeness, and quality of reflection. Use the rubric on page 158.	For the class: • transparencies (optional)
• **Individual Self-Assessment Rubric:** Students can use the checklist on page 106 to determine how well they met behavior and work expectations.	• overhead projector (optional) • transparency markers (optional)
• **Team Evaluation:** Students can complete the questions on page 107 to reflect on how well they worked in teams and celebrate successes, as well as make plans to improve teamwork.	
• **Student Participation Rubric:** Make copies of the rubric on page 159 to score each student's participation in the design challenge.	

	Assessment	Materials
DESIGN CHALLENGE 3: OUTBREAK!	• **2. Research:** Assess how well students can recognize and describe patterns from a sequence of numbers or from its graph. • **2. Research:** Assess how well students can generalize the pattern using symbols. • **2. Research:** Assess how well students can extrapolate data using the patterns they found. • **2. Research:** Assess whether students can apply the findings of their research to create a rule of thumb to help meet design criteria. • **6–8. Test, Communicate, Redesign:** Assess written responses and student observations during test, communicate, and redesign steps based on model performance, completeness, and quality of reflection. Use the rubric on page 158. • **Individual Self-Assessment Rubric:** Students can use the checklist on page 145 to determine how well they met behavior and work expectations. • **Team Evaluation:** Students can complete the questions on page 146 to reflect on how well they worked in teams and celebrate successes as well as make plans to improve teamwork. • **Student Participation Rubric:** Make copies of the rubric on page 159 to score each student's participation in the design challenge.	For each team: • 1 calculator • 1 pack of colored pencils (at least 3 colors) • 1 penny • access to a computer For the teacher: • 1 lump of clay or modeling compound • 1 CD of the virus-simulation applets • 1 LCD projector and computer For the class: • sheet of chart paper

AMAZON MISSION MASTER MATERIALS LIST

Qty	Item	C* or R*	1. Malaria Meltdown!	2. Mercury Rising!	3. Outbreak!
PER GROUP					
1–2	calculator(s)	R	✓	✓	✓
1	chopstick	R		✓	
1	digital thermometer	R	✓		
1	large nail	R		✓	
1	measuring cup with 250-mL line	R		✓	
1	meterstick	R	✓		
1	pack of colored pencils (at least 3 colors)	R	✓		✓
1	pack of transparency markers (4 colors)	R	✓	✓	
1–2	pair(s) of scissors	R	✓	✓	
1	penny	R			✓
1	pin	R		✓	
90	plastic spheres (diameter 1.2 cm)[1]	R		✓	
1	ruler (metric)	R		✓	
1	small nail	R		✓	
1	stopwatch	R	✓	✓	
1	water basin or pail	R		✓	
1	bottle of water (about 500 mL)	C		✓	
1	cup of ice	C	✓		
1	egg (raw)	C	✓		
4	foam cups (12-oz)	C		✓	
9	foam cups (8-oz)	C		✓	
1	piece of aluminum foil (about 2 m²)	C	✓		
1	piece of bubble wrap (about 2 m²)	C	✓		
1	piece of cardstock (standard letter size)	C	✓		
1	piece of corrugated cardboard (about 3 m²)	C	✓		
1	piece of foam board (about 3 m²)	C	✓		
2	plastic bags (small)	C		✓	
6	plastic stirrers	C		✓	
6	plastic straws	C		✓	
1	resealable sandwich bag	C	✓		
2	rolls of packing tape (heavy-duty)	C	✓	✓	
3	transparencies	C	✓	✓	
PER TEACHER					
1	DVD of virus-simulation applets	R			✓
1	large lump of clay or modeling compound	R			✓
1	LCD projector and computer	R			✓
1	overhead projector	R	✓	✓	
1	piece of chart paper	C	✓	✓	✓

[1]Mardi Gras beads work well as plastic spheres.   *C = Consumable   *R = Reusable

COMMON CORE AND ITEEA STANDARDS CORRELATIONS

The following tables show how each design challenge addresses Common Core State Mathematics Standards and International Technology and Engineering Standards. In the Common Core column, double asterisks (**) denote standards that are not expressly addressed by the design challenges, but that can be addressed by using optional suggestions included in the instructional text for that design challenge. References to the specific pages are included.

	Common Core State Standards for Mathematics (Grades 6–8)[1]	ITEEA Standards for Technological Literacy (STL)[2]
DESIGN CHALLENGE 1: MALARIA MELTDOWN!	**Mathematical Practices** 2. Reason abstractly and quantitatively.** **See Optional CCSS Enhancement(s) on page 49. 5. Use appropriate tools strategically. 6. Attend to precision. **Standards** **6.EE.9.** Use variables to represent two quantities in a real-world problem that change in relationship to one another; write an equation to express one quantity, thought of as the dependent variable, in terms of the other quantity, thought of as the independent variable. Analyze the relationship between the dependent and independent variables using graphs and tables, and relate these to the equation. **6.G.4.** Represent three-dimensional figures using nets made up of rectangles and triangles, and use the nets to find the surface area of these figures. Apply these techniques in the context of solving real-world and mathematical problems. **7.RP.2.a.** Decide whether two quantities are in a proportional relationship, e.g., by testing for equivalent ratios in a table…. **7.RP.2.b.** Identify the constant of proportionality (unit rate) in tables … equations, diagrams, and verbal descriptions of proportional relationships. **7.RP.2.c.** Represent proportional relationships by equations.** **See Optional CCSS Enhancement(s) on page 57.	**1F** New products and systems can be developed to solve problems or to help do things that could not be done without the help of technology. **1G** The development of technology is a human activity and is the result of individual and collective needs and the ability to be creative. **1H** Technology is closely linked to creativity, which has resulted in innovation. **2R** Requirements are the parameters placed on the development of a product or system. **2S** Trade-off is a decision process recognizing the need for careful compromises among competing factors. **8E** Design is a creative planning process that leads to useful products and systems. **8F** There is no perfect design. **8G** Requirements for design are made up of criteria and constraints. **9F** Design involves a set of steps, which can be performed in different sequences and repeated as needed. **9G** Brainstorming is a group problem-solving design process in which each person in the group presents his or her ideas in an open forum. **9H** Modeling, testing, evaluating, and modifying are used to transform ideas into practical solutions. **11H** Apply a design process to solve problems in and beyond the laboratory-classroom.

[1]Common Core State Standards. Copyright 2010. National Governor's Association Center for Best Practices and Council of Chief State School Officers. All rights reserved.

[2]International Technology and Engineering Educators Association (ITEEA). 2007. *Standards for Technological Literacy: Content for the Study of Technology.* (Third ed.) Virginia.

COMMON CORE AND ITEEA STANDARDS CORRELATIONS
(CONTINUED)

	Common Core State Standards for Mathematics (Grades 6–8)	ITEEA Standards for Technological Literacy (STL)
DESIGN CHALLENGE 1: MALARIA MELTDOWN!	**7.G.6.** Solve real-world and mathematical problems involving area, volume, and surface area of two- and three-dimensional objects composed of triangles, quadrilaterals, polygons, cubes, and right prisms. **8.EE.5.** Graph proportional relationships, interpreting the unit rate as the slope of the graph. Compare two different proportional relationships represented in different ways. **8.F.5.** Describe qualitatively the functional relationship between two quantities by analyzing a graph (e.g., where the function is increasing or decreasing, linear or nonlinear). Sketch a graph that exhibits the qualitative features of a function that has been described verbally.	**11J** Make two-dimensional and three-dimensional representations of the designed solution. **11K** Test and evaluate the design in relation to pre-established requirements, such as criteria and constraints, and refine as needed. **11L** Make a product or system and document the solution.

DESIGN CHALLENGE 2: MERCURY RISING!

Common Core State Standards for Mathematics (Grades 6–8)	ITEEA Standards for Technological Literacy (STL)

Mathematical Practices
3. Construct viable arguments and critique the reasoning of others.
4. Model with mathematics.
5. Use appropriate tools strategically.
 	
Standards
6.RP.3. Use ratio and rate reasoning to solve real-world and mathematical problems, e.g., by reasoning about tables of equivalent ratios, … double number line diagrams,** or equations.
***See Optional CCSS Enhancement(s) on page 86.*
   a. Make tables of equivalent ratios relating quantities with whole-number measurements, find missing values in the tables, and plot the pairs of values on the coordinate plane. Use tables to compare ratios.
   b. Solve unit rate problems including those involving unit pricing and constant speed.
   c. Find a percent of a quantity as a rate per 100 (e.g., 30% of a quantity means 30/100 times the quantity); solve problems involving finding the whole, given a part and the percent.**
***See Optional CCSS Enhancement(s) on page 78.*
6.EE.2. Write, read, and evaluate expressions in which letters stand for numbers.**
***See Optional CCSS Enhancement(s) on page 78.*
   a. Write expressions that record operations with numbers and with letters standing for numbers. **
***See Optional CCSS Enhancement(s) on page 78.*
   b. Identify parts of an expression using mathematical terms (sum, term, product, factor, quotient, coefficient); view one or more parts of an expression as a single entity.**
***See Optional CCSS Enhancement(s) on page 78.*
   c. Evaluate expressions at specific values of their variables. Include expressions that arise from formulas used in real-world problems. Perform arithmetic operations, including those involving whole-number exponents, in the conventional order when there are no parentheses to specify a particular order (Order of Operations).
	1F New products and systems can be developed to solve problems or to help do things that could not be done without the help of technology.
	
	1G The development of technology is a human activity and is the result of individual and collective needs and the ability to be creative.
	
	1H Technology is closely linked to creativity, which has resulted in innovation.
	
	2R Requirements are the parameters placed on the development of a product or system.
	
	2S Trade-off is a decision process recognizing the need for careful compromises among competing factors.
	
	8E Design is a creative planning process that leads to useful products and systems.
	
	8F There is no perfect design.
	
	8G Requirements for design are made up of criteria and constraints.
	
	9F Design involves a set of steps, which can be performed in different sequences and repeated as needed.
	
	9G Brainstorming is a group problem-solving design process in which each person in the group presents his or her ideas in an open forum.
	
	9H Modeling, testing, evaluating, and modifying are used to transform ideas into practical solutions.

COMMON CORE AND ITEEA STANDARDS CORRELATIONS
(CONTINUED)

Common Core State Standards for Mathematics (Grades 6–8)	ITEEA Standards for Technological Literacy (STL)
7.RP.1. Compute unit rates associated with ratios of fractions, including ratios of lengths, areas, and other quantities measured in like or different units. **7.EE.4.a.** Solve word problems leading to equations of the form $px + q = r$ and $p(x + q) = r$, where p, q, and r are specific rational numbers. Solve equations of these forms fluently. Compare an algebraic solution to an arithmetic solution, identifying the sequence of the operations used in each approach. **7.EE.4.b.** Solve word problems leading to inequalities** of the form $px + q > r$ or $px + q < r$, where p, q, and r are specific rational numbers. Graph the solution set of the inequality and interpret it in the context of the problem. ***See Optional CCSS Enhancement(s) on page 75.* **8.EE.5.** Graph proportional relationships, interpreting the unit rate as the slope of the graph. Compare two different proportional relationships represented in different ways. **8.G.9.** Know the formulas for the volumes of … spheres** and use them to solve real-world and mathematical problems. ***See Optional CCSS Enhancement(s) on page 78.*	**11H** Apply a design process to solve problems in and beyond the laboratory-classroom. **11J** Make two-dimensional and three-dimensional representations of the designed solution. **11K** Test and evaluate the design in relation to pre-established requirements, such as criteria and constraints, and refine as needed. **11L** Make a product or system and document the solution.

DESIGN CHALLENGE 2: MERCURY RISING!

COMMON CORE AND ITEEA STANDARDS CORRELATIONS
(CONTINUED)

DESIGN CHALLENGE 3: OUTBREAK!

Common Core State Standards for Mathematics (Grades 6–8)	ITEEA Standards for Technological Literacy (STL)

Mathematical Practices
3. Construct viable arguments and critique the reasoning of others.
4. Model with mathematics.
7. Look for and make use of structure.
 	
Standards
6.EE.1. Write and evaluate numerical expressions involving whole-number exponents.
6.SP.5. Summarize numerical data sets in relation to their context, such as by:
  a. Reporting the number of observations.
  b. Describing the nature of the attribute under investigation, including how it was measured and its units of measurement.
  c. Giving quantitative measures of center (median and/or mean**) … as well as describing any overall pattern and any striking deviations from the overall pattern with reference to the context in which the data were gathered.
  **See Optional CCSS Enhancement(s) on page 110.*
  d. Relating the choice of measures of center and variability** to the shape of the data distribution** and the context in which the data were gathered.
  **See Optional CCSS Enhancement(s) on page 110.*
7.EE.3. Solve multi-step real-life and mathematical problems posed with positive and negative rational numbers in any form (whole numbers, fractions, and decimals), using tools strategically. Apply properties of operations to calculate with numbers in any form; convert between forms as appropriate; and assess the reasonableness of answers using mental computation and estimation strategies.
7.SP.5. Understand that the probability of a chance event is a number between 0 and 1 that expresses the likelihood of the event occurring. Larger numbers indicate greater likelihood. A probability near 0 indicates an unlikely event, a probability around 1/2 indicates an event that is neither unlikely nor likely, and a probability near 1 indicates a likely event.**
  **See Optional CCSS Enhancement(s) on page 123.*
	1F New products and systems can be developed to solve problems or to help do things that could not be done without the help of technology.
	
	1G The development of technology is a human activity and is the result of individual and collective needs and the ability to be creative.
	
	1H Technology is closely linked to creativity, which has resulted in innovation.
	
	2R Requirements are the parameters placed on the development of a product or system.
	
	2S Trade-off is a decision process recognizing the need for careful compromises among competing factors.
	
	8E Design is a creative planning process that leads to useful products and systems.
	
	8F There is no perfect design.
	
	8G Requirements for design are made up of criteria and constraints.
	
	9F Design involves a set of steps, which can be performed in different sequences and repeated as needed.
	
	9G Brainstorming is a group problem-solving design process in which each person in the group presents his or her ideas in an open forum.
	
	9H Modeling, testing, evaluating, and modifying are used to transform ideas into practical solutions.

Common Core State Standards for Mathematics (Grades 6–8)	ITEEA Standards for Technological Literacy (STL)
8.F.5. Describe qualitatively the functional relationship between two quantities by analyzing a graph (e.g., where the function is increasing or decreasing, linear or nonlinear). Sketch a graph that exhibits the qualitative features of a function that has been described verbally. **8.SP.4.** Understand that patterns of association can also be seen in bivariate categorical data by displaying frequencies and relative frequencies in a two-way table. Construct and interpret a two-way table summarizing data on two categorical variables collected from the same subjects. Use relative frequencies calculated for rows or columns to describe possible association between the two variables.	**11H** Apply a design process to solve problems in and beyond the laboratory-classroom. **11K** Test and evaluate the design in relation to pre-established requirements, such as criteria and constraints, and refine as needed. **11L** Make a product or system and document the solution.

DESIGN CHALLENGE 3: OUTBREAK!

PACING PLANNING GUIDE

*ESTIMATED TIME ASSUMES CLASS PERIODS OF 45 TO 50 MINUTES.

Each design challenge will take 5 or more days.

Section name	EDP step[1]	Page number	Estimated time*	Your time estimate
Team-Building Activity (optional)		1	Day 0	
Amazon Mission Prerequisite Math Skills (optional)		3	Day 0	
Amazon Mission Introduction		22	Day 1	
Introducing the Engineering Design Process (EDP)		26	Day 1	
DESIGN CHALLENGE 1: MALARIA MELTDOWN!				
Define: Design criteria and constraints are defined.	1	34	Day 1	
Research: Students interpret a temperature versus time graph that shows the performance of an existing medicine carrier, and answer questions.	2	36	Day 2	
Research: Students experiment to see how well one layer of each material insulates; they collect and graph data, and interpret the graph.	2	42	Day 3	
Research: Students experiment to see how well combinations of two materials insulate; they collect and graph data; interpret the graph; and present results to the class.	2	48	Days 3–4	
Brainstorm: Given the cost of materials, students individually sketch medicine carrier design.	3	52	Day 4, homework	
Choose: Team decides on carrier design; students sketch a 3-D drawing and a 2-D net, and calculate the surface area to determine cost of materials.	4	56	Day 5	
Build: Students build a prototype of the carrier.	5	61	Day 5	
Test: Students test carrier prototype.	6	63	Day 6	
Communicate: Students answer questions and present to the class.	7	65	Days 6–7	
Redesign: Students answer questions about improving design.	8	67	Day 7	

[1]To learn more about the Engineering Design Process (EDP), see pages 148–149.

Section name	EDP Step	Page number	Estimated time	Your time estimate
DESIGN CHALLENGE 2: MERCURY RISING!				
Define: Design criteria and constraints are defined.	1	75	Day 1	
Research: Students calculate surface area of spheres to figure out which size sphere would meet criteria and be lowest in cost.	2	77	Day 1	
Research: Students convert units to find out minimum and maximum time for filtering 250 mL of water.	2	81	Day 2	
Research: Students experiment with flow rate through different size holes; they collect and graph data, and interpret the graph.	2	83	Day 2	
Research: Students design their own experiments to investigate other factors that might affect flow rate of filter; they collect and graph data, interpret the graph, and present results to the class.	2	85	Day 3	
Brainstorm: Students individually sketch filter design.	3	92	Days 3–4, homework	
Choose: Team decides on filter design and draws design.	4	95	Day 4	
Build: Students build scale prototype of filter design.	5	97	Day 4	
Test: Students test prototype to see if criteria are met.	6	99	Day 5	
Communicate: Students answer questions and share with the whole class.	7	101	Days 5–6	
Redesign: After hearing about other teams' designs, students answer questions to improve their own designs.	8	103	Day 6	

Section name	EDP step	Page number	Estimated time	Your time estimate
DESIGN CHALLENGE 3: OUTBREAK!				
Define: Design criteria/constraints are defined.	1	110	Day 1	
Research: Students simulate the spread of a contagious virus (with 100% infection rate), and describe and generalize exponential patterns.	2	112	Day 1	
Research: Students simulate the spread of a virus with a doctor who can cure 1 person per time interval; students generalize patterns using symbols.	2	115	Day 1	
Research: Students simulate the spread of a virus with air filtration masks that reduce the infection rate by 50%, and describe patterns.	2	118	Day 2	
Research: Students graph simulation data and compare results.	2	121	Day 2	
Research: Students calculate probabilities of combinations of interventions.	2	123	Day 2	
Research (*optional*): Students use virus-simulation applet to find the lowest cost and working plan that only uses doctors.	2	125	Day 3	
Brainstorm: Students brainstorm virus intervention plans individually.	3	129	Days 3–4, homework	
Choose: Team members discuss benefits and drawbacks of individual plans; team chooses best plan and answers questions.	4	132	Day 4	
Build: Teacher shows how to adjust variables in computer applet according to chosen virus intervention plan; teams enter plans into applet.	5	134	Day 4	
Test: Students run simulation and record results, then see if criteria are met; teams are challenged to tweak plan to find the lowest cost plan that meets the virus-containment criterion.	6	136	Day 4	
Communicate: Students answer questions and share with the class.	7	140	Days 4–5	
Redesign: After hearing from other teams, students answer questions to compare designs and improve their own designs.	8	142	Day 5	

Teacher Page

"THE LAW FOR THE WOLVES" TEAM-BUILDING ACTIVITY *(OPTIONAL)*

Students work and communicate in teams during most of each design challenge. Some pilot teachers found it useful to do some team-building activities prior to the start of the unit. There is a different team-building activity in each of the *Building Math* books.

OBJECTIVE:

- Students read and discuss a poem about teamwork.

GROUP SIZE:

- whole class

SETUP:

- Make a copy of the poem excerpt on the next page for each student.
  OR
- Make a copy of the poem on a transparency to project on the board or wall with an overhead projector.
  OR
- Project the poem on the board or wall using a computer/LCD projector.

PROCEDURES:

1. Distribute copies of the poem or project the poem on the board or wall.

2. Read the poem together and discuss the following:

   - Why does the poet describe the law as "runneth forward and back"?
   - Why are there two parts to this law? How are they different?
   - Why is this law essential to the wolf's survival?
   - How is this poem related to teamwork?

3. Summarize: Each person has unique gifts and abilities to contribute to the group. Not only is the group strengthened because of your unique contribution, but you also grow as a person from being part of the group. You can often do more as a group than as an individual. Working together can bring out the best in each of you. Teamwork is not easy, though, because you will have to learn to share, negotiate, listen, and compromise; you may not always get your way.

THE LAW FOR THE WOLVES
(ABRIDGED)

Rudyard Kipling

Now this is the law of the jungle,
as old and as true as the sky,
And the wolf that shall keep it may prosper,
but the wolf that shall break it must die.

As the creeper that girdles the tree trunk,
the law runneth forward and back;
For the strength of the pack is the wolf,
and the strength of the wolf is the pack.

Teacher Page

AMAZON MISSION PREREQUISITE MATH SKILLS (*OPTIONAL*)

STUDENTS SHOULD BE ABLE TO			
MATH SKILL	1. MALARIA MELTDOWN!	2. MERCURY RISING!	3. OUTBREAK!
Make a line graph.	✓	✓	✓
Find the slope of two points on a line or curve.	✓		
List combinations.	✓		
Calculate the surface area of prisms, cylinders, and other 3-D objects.	✓		
Draw the net of common 3-D objects such as prisms, cylinders, cones, and so forth.	✓		
Use a formula to calculate surface area.	✓		
Represent lower and upper constraints as a compound inequality.	✓	✓	
Measure length (in metric units) using a ruler.	✓	✓	
Convert units within a measurement system and use proportional reasoning.		✓	
Recognize exponential patterns and represent them using symbols or variables as direct and recursive equations.			✓
Calculate compound probabilities.			✓

* Bold text indicates key objectives.

HOW TO USE THE WORKSHEETS

Use the following tips for review and reinforcement as needed.

1. Line Graph Activity (pages 5–9)

    - The goal of this activity is for students to:
        - make a line graph
        - identify independent and dependent variables
        - use convention to put the independent variable on the *x*-axis and the dependent variable on the *y*-axis
        - use range of data to set up scales on axes so that the data is well spread out
        - use equal intervals when setting up scales on axes
        - label the axes with data type and unit
        - label the graph with an appropriate title

    - Use Exercise 1 to guide students through the steps of constructing a line graph—particularly steps 2 and 3 (scaling the axes).

    - If students need additional guidance in scaling the axes, use the Line Graph Scaling Examples on pages 10–13.

    - Assign students to do Exercise 2 on their own or with a partner.

2. Converting Units Activity (pages 14–16)

    The goal of this activity is for students to convert units within a measurement system and use proportional reasoning.

    - Before distributing the problems, go through the activity with students.

    - Assign students to work through the problems on their own or with a partner.

3. Representing Patterns Activity (pages 17–20)

    - The goal of this activity is for students to recognize exponential and linear patterns, and represent them using symbols or variables as direct and recursive equations.

    - Go over Example 1 with students.

    - Assign students to work through the problems on their own or with a partner.

LINE GRAPH ACTIVITY

A line graph is a way to visually show how two sets of data are related and how they vary depending on each other. Line graphs are particularly effective for showing change over time, predicting what comes next, or estimating what happened in between data points.

HOW TO CONSTRUCT A LINE GRAPH ON PAPER

STEP	WHAT TO DO	HOW TO DO IT
1	Identify the variables.	a. Ask yourself, "Which of the variables did I control or vary?" These values (independent variables) are the ones that you measure or choose before conducting the experiment. b. The *x*-axis (horizontal) typically represents the independent variable. c. Ask yourself, "Which of the variables was affected as a result of the experiment?" These values (dependent variables) are the ones that you measure during the experiment that correlate one-on-one with the independent variable values. d. The *y*-axis (vertical) typically represents the dependent variable.

2	Determine a scale for each axis.	a. Your goal in determining a scale for the axes is to fit the entire range of data over the available space on the graph paper.
		
		b. Find the range of the data values (lowest to highest values). If necessary, round down the lowest data value and round up the highest data value to the nearest whole number or power of 10. Find the difference.
		
		c. Count the squares on the axis that you want to use to represent the range. It is fine to round down. If your data doesn't start at 0, subtract 1 from the number of squares.
		
		d. Determine the scale: data range to total number of boxes along the axis.
		
		  i. If total number of boxes is greater than the data range:
		    • (total number of boxes) ÷ range = a
		    • a represents the number of boxes in between each whole interval value
		    • round UP if necessary
		  ii. If total number of boxes is less than the range:
		    • range ÷ (number of boxes) = b
		    • b represents the interval value of each box
		    • round UP if necessary

3	Number and label each axis.	a. Mark the scale values on each axis. b. Label each axis with the type of data and unit.
4	Plot the data points.	a. Plot each pair of data values (independent-dependent pair) on the graph with a dot. b. You may have multiple sets of dependent-variable data. If so, use a different color to plot each set of data pairs.

5	Draw the graph.	Draw a curve or a line that best fits the data points.
6	If necessary, include a key.	If you have multiple sets of dependent-variable data, use a color-coded key to name each set.
7	Title the graph.	a. Your title should include both of the variables that are being compared. b. Someone should be able to read the title and know exactly what the graph shows.

EXERCISE 1

The data table below shows the average value of a truck as the mileage on the truck increases. Answer the questions that follow and make a line graph to represent the data.

MILEAGE (KILOMETERS)	0	20,000	40,000	60,000	80,000	100,000	120,000
TRUCK'S VALUE (DOLLARS)	$14,000	$12,000	$8,000	$5,000	$4,000	$3,500	$3,000

1. What is the independent variable? How do you know?

2. What is the dependent variable? How do you know?

3. Write x next to the data table row that contains data values for the x-axis.

4. Write y next to the data table row that contains data values for the y-axis.

5. To scale the x-axis:

    a. Find the range of the data values: _____ to _____. The difference is: _____.

    b. Count the number of boxes along the x-axis: _____.

    c. Determine the scale: data range _____ to total number of boxes _____.

        i. If data range is greater than the total number of boxes, calculate data range ÷ total number of boxes = _____.

        ii. If data range is less than the total number of boxes, calculate (number of boxes) ÷ (data range) = _____.

    d. Use the above information to label the scale values.

    e. Label the x-axis.

6. Repeat step 5 to scale the *y*-axis. Show your work below.

7. Plot the data on the graph.

8. Give the graph a title that best describes the data shown.

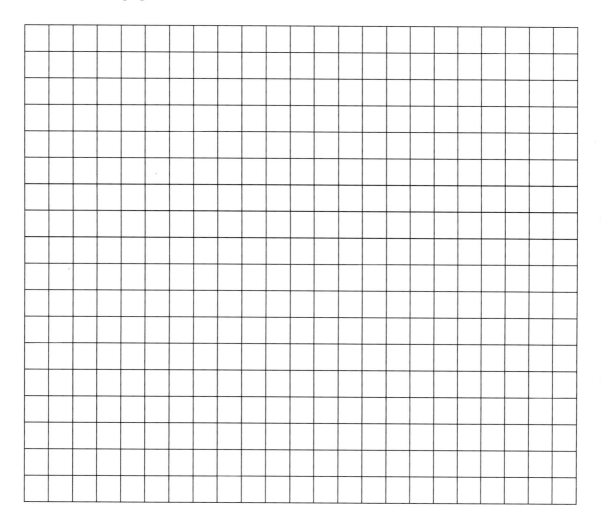

Use the graph to answer the questions below.

9. What is the value of the truck when the mileage is 100,000?

10. When does the value of the truck decrease the most?

11. About how much is the truck worth when the mileage is 50,000?

12. How much will the truck be worth at 140,000 miles? How do you know?

EXERCISE 2

The data table below shows the average thickness of annual tree-ring growth in two forests as the trees age. Assume that thicknesses of the tree rings were measured during the same years. A thin ring usually indicates lack of water, forest fires, or a major insect infestation. A thick ring indicates just the opposite. Make a graph of the data.

AGE OF THE TREE (YEARS)	AVERAGE THICKNESS OF THE ANNUAL RINGS IN FOREST A (CM)	AVERAGE THICKNESS OF THE ANNUAL RINGS IN FOREST B (CM)
10	2.0	2.2
20	2.2	2.5
30	3.5	3.6
35	3.0	3.8
50	4.5	4.0
60	4.3	4.5

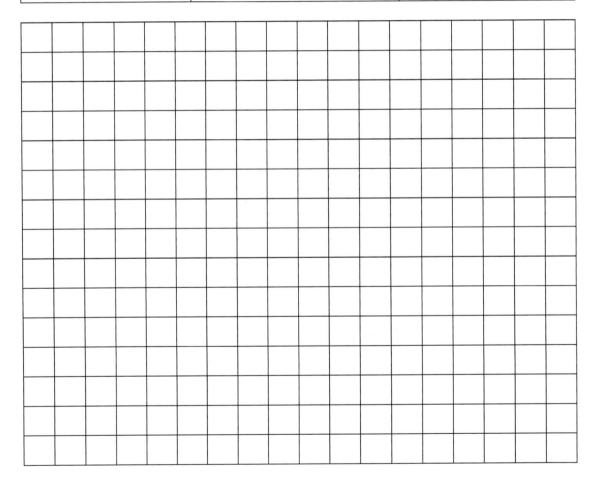

Use the graph to answer the questions below.

1. What was the average thickness of annual rings of 20-year-old trees in Forest *B*?

2. How old were the trees in both forests when the average thicknesses of the annual rings were about the same?

3. What would be a reasonable prediction for the average thickness of annual rings of 70-year-old trees in Forest *B*?

4. Based on this data, what can you conclude about Forest *A* and Forest *B*? How do you know? What is your evidence?

Teacher Page

LINE GRAPH SCALING EXAMPLES

Students probably struggle the most with choosing appropriate scales for the axes that fit the full range of data on the available graph paper space. Below are two detailed step-by-step examples of how to scale different kinds of data sets—ones that start with zero and ones that don't, ones that contain decimal values, and ones that contain wide and narrow ranges. Use these to provide direct instruction, as necessary.

SCALING EXAMPLE 1 (DATA STARTS AT 0)

The data table shows the time and distance a car traveled on a road. Make a line graph of the data table.

TIME (HR)	0	1	2	3	4	5	6
DISTANCE (KM)	0	50	100	150	200	230	250

1. The time variable is the independent variable. It'll go on the *x*-axis. The distance variable is the dependent variable. It'll go on the *y*-axis.

2. What is the range of the time data? 0 to 6. The difference is 6.

3. Count the number of boxes on the *x*-axis: 18.

4. You have 18 boxes to represent 6 hours. Since you have more boxes than hours, you'll need multiple boxes to represent each hour. How many boxes would represent 1 hour? Divide the number of boxes by the number of equal intervals: 18 ÷ 6 = 3. This means that 3 boxes represent 1 hour.

5. On the *x*-axis, starting at the left-most line and labeling that 0, count 3 boxes and then draw a tick mark and call it 1 hour. Continue counting 3 boxes, drawing tick marks, and labeling each tick mark.

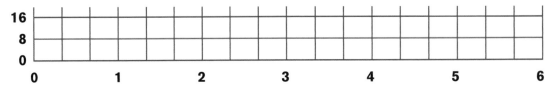

6. What is the range of the distance data? 0 to 250. The difference is 250. Count the number of boxes on the *y*-axis: 32.

7. You have 32 boxes to represent 250 kilometers. Since you have fewer boxes than kilometers, each box needs to represent multiple kilometers. So how many kilometers would 1 box represent? Divide the number of kilometers by the number of boxes: 250 ÷ 32 = 7.8125. This means that 1 box would represent 7.8125 kilometers. If you round up to 8 kilometers, 32 boxes would represent more than the necessary 250 range because 32 × 8 = 256. But if you round down to 7 kilometers = 1 box, you will only cover a range of 32 × 7 = 224, which isn't enough.

    1 box = 8 kilometers

    10 boxes = 80 kilometers

    5 boxes = 40 kilometers

© Museum of Science (Boston), Wong, Brizuela

8. On the *y*-axis, starting at the bottom most line and labeling that 0, you can mark the next box 8 (0 + 8 = 8) and continue marking each box 8 more than the previous one. You can also count in increments of 5 boxes and add 40, or count in increments of 10 boxes and add 80. See below. What are the advantages and disadvantages of each kind of labeling?

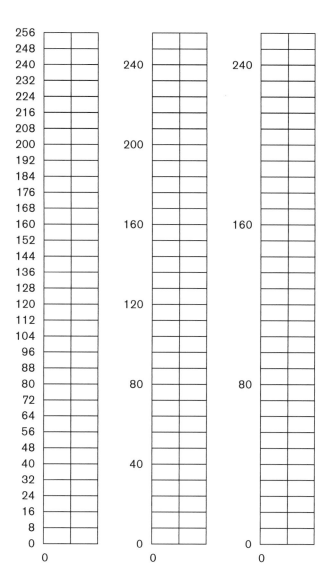

9. Label your axes and give the graph a title.

SCALING EXAMPLE 2 (DATA DOES NOT START AT 0)

The data table below shows the measurements of thigh lengths of runners and the runners' corresponding times in a 100-meter dash.

THIGH LENGTH (CM)	24	31	37	38	39	42	51	55	62	71
TIME OF 100-METER DASH (SEC)	24.6	25.2	29.9	27.3	22.4	23.0	22.1	25.4	24.6	27.3

1. The independent variable is the thigh length, and the dependent variable is the time of the 100-meter dash. This means that the thigh length values will go on the *x*-axis, and the time of the 100-meter dash will go on the *y*-axis.

2. Range of thigh length is 24 to 71. Round 24 to 20 and 71 to 75 so the approximate range is 20 to 75, for a difference of 55. Why is it okay to round the lowest value down and the highest value up but NOT vice versa? It's okay (even recommended) because your goal in creating this scale is to fit the range of data. Having a range of 55 would definitely fit all your data points, with some room below and above the range. But it's not okay to round the lowest value up and the highest value down because then you won't cover the full range of data values.

3. The number of boxes on the *x*-axis is 23. Round this down to 20 to make it easier to divide. Why is it okay to round down but not okay to round up? It's okay to round down because you don't need to use all the boxes to represent the data. It's not okay to round up because you only have 23 boxes on the axis. If you say you have more, then you will need to draw more boxes to fit the range of data.

    Note: Even if you don't need to round down, subtract at least 1 box from the maximum number of boxes. This is because you're not starting at 0 and would need to start at least 1 box away from the first line.

4. Scale: data range 55 to 20 boxes.

5. Since the data range is higher than the number of boxes, 55 (data range) ÷ 20 (number of boxes) = 2.75. Round this up to 3 so each box represents 3 cm.

6. Since the data doesn't start at 0, draw a squiggly line between the first line and the second line of the *x*-axis to indicate a break in the data values. This means that the first box does not represent the same data interval as the others to follow.

7. Label the second line with the lowest value of your APPROXIMATED range (rounded from 24 to 20). So starting at 20, count 1 box and label the next line whatever the previous value is plus 3. So your scale would be 20, 23, 26, 29, 32, 35, and so forth, with 1 box in between each value. You should have no problem reaching the approximated upper range value of 75 without running out of boxes.

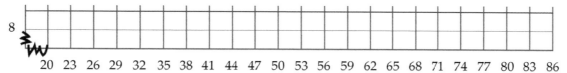

8. Range of 100-meter dash time is 22.1 to 29.9. Round 22.1 down to 22 and 29.9 up to 30 so the approximated range is 22 to 30, for a difference of 8.

9. The number of boxes on the graph is 32.

10. Scale: data range 8 to 32 boxes.

11. Since the number of boxes is higher than the data range, 32 (number of boxes) ÷ 8 (data range) = 4 boxes per 1 second. Furthermore, since the data range includes decimal numbers with digits in the tenth places, each box can represent 0.25 second.

12. Again, since the data doesn't start at 0, draw a squiggly line between the first line and the second line of the *y*-axis to indicate a break in the data values.

13. Label the second line with the lowest value of the approximated range, which is 22, rounded down from 22.1. Starting from 22, count 4 boxes, mark 23, and so forth. Since there are 4 boxes in between each whole number value, each box represents 0.25 second.

14. Label the axes and give the graph a title.

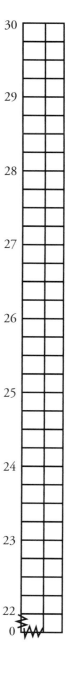

Teacher Page

CONVERTING UNITS ACTIVITY

ACTIVITY INSTRUCTION

1. Ask the class, "Who's a fast walker or runner?" Invite different students to guess how many miles they can walk or run per hour.

2. Choose one student's speed (e.g., 5 km per hour) and use it as the example. Say to the class, "Let's say we want [student's name] to show us that speed in this class. We don't have the time to watch [student's name] run for an hour, nor do we have the space in this classroom. But we want to see that speed demonstrated from one side of the classroom to the other."

3. Measure the classroom length from one side to another. Let's say it's 10 meters.

4. Say to the class, "So we want to convert 5 kilometers per hour to 10 meters per x seconds so [student's name] can show us what that 5 kilometers per hour looks like when walking or running only 10 meters instead of 5 kilometers."

5. Write the following equation on the board:

$$\frac{5 \text{ km}}{1 \text{ hour}} = \frac{10 \text{ m}}{x \text{ seconds}}$$

6. One way to convert units is to use multipliers. Ask the class, "How many meters are in 1 kilometer?" Answer: There are 1,000 meters in a kilometer. So to convert 5 kilometers to meters, we use a multiplier as shown below:

$$\frac{5 \text{ km}}{1 \text{ hour}} \cdot \frac{1,000 \text{ m}}{1 \text{ km}} = \frac{5,000 \text{ m}}{1 \text{ hour}}$$

7. Emphasize that multipliers are set up in a way so that the units you want to cancel are always diagonal to each other. In this case, we want to convert from kilometers to meters.

8. Next, convert hours to seconds. Usually, we think of minutes in an hour and then seconds in a minute, so we can use two multipliers as shown below:

$$\frac{5,000 \text{ m}}{1 \text{ hour}} \cdot \frac{1 \text{ hour}}{60 \text{ minutes}} \cdot \frac{1 \text{ minute}}{60 \text{ seconds}} = \frac{5,000 \text{ m}}{3,600 \text{ seconds}}$$

9. So now we know that 5 km per hour is equivalent to 5,000 meters per 3,600 seconds.

10. However, we need to know how many seconds it would take the student to run or walk 10 meters across the classroom at this rate, so we need to use some proportional reasoning.

$$\frac{5,000 \text{ meters}}{3,600 \text{ seconds}} = \frac{10 \text{ meters}}{x \text{ seconds}}$$

11.   There are several ways to solve this proportion. One way is to think about the proportion as two equivalent fractions and use the observation that 10 meters can divide evenly into 5,000 m exactly 500 times. Divide the denominator by the same amount to get the value of the unknown quantity:

$$\frac{5,000 \text{ m}}{3,600 \text{ seconds}} \begin{array}{c} \div\ 500 \\ \div\ 500 \end{array} \approx \frac{10 \text{ meters}}{7.2 \text{ seconds}}$$

12.   Thus, it would take [student's name] 7.2 seconds to run 10 meters if he or she is going at 5 kilometers per hour.

13.   Another way to solve the proportion is to cross-multiply and solve it as an equation:

$$\frac{5,000 \text{ m}}{3,600 \text{ seconds}} = \frac{10 \text{ meters}}{x \text{ seconds}}$$

$$5,000x = 10 \cdot 3,600$$

$$5,000x = 36,000$$

$$\frac{5,000x}{5,000} = \frac{36,000}{5,000}$$

$$x = 7.2$$

$$x \approx 7.2 \text{ seconds}$$

14.   Now you can ask the student to demonstrate running the 10 meters in the classroom or hallway in 7.2 seconds. Use a stopwatch.

CONVERTING UNITS ACTIVITY

1. What is 10 kilometers per hour converted to 10 meters per x seconds?

2. What is 5 kilometers per hour converted to 5 meters per x seconds?

3. What is 200 kilometers per day converted to 50 meters per x minutes?

4. What is 25 liters per day converted to 100 milliliters (mL) per x seconds?

REPRESENTING PATTERNS ACTIVITY

EXAMPLE

Below is a common test question in which you are asked to recognize, extend, and represent patterns.

A worker used black and white tiles to make the following pattern:

 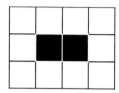 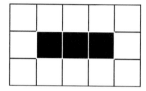

a. Based on the pattern, how many white tiles would be needed for a figure with 4 black tiles?

To solve the problem, you may simply draw the next figure in the sequence and count the number of tiles. You can also make a table and discover a regularly increasing pattern in the numbers.

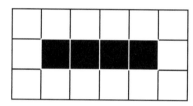

b. What patterns do you notice in the numbers of black and white tiles in the figures as the pattern continues? How can you represent these patterns using words or symbols?

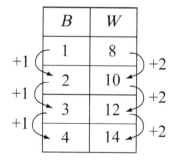

You may see that as the pattern continues, the number of black tiles increases by 1 and the number of white tiles increases by 2. The pattern may be represented as follows:

$B_{next} = B_{current} + 1$

$W_{next} = W_{current} + 2$

These equations are called "recursive," because they can only predict the next number in the sequence if the previous number is known. These patterns are fairly easy to recognize, even when the patterns are exponential (for example, 1, 2, 4, 8, 16, 32, and so forth).

c. Based on this pattern, how many white tiles would be needed for a figure with 50 black tiles?

You could use your recursive equation to continue the pattern to find the number of white tiles that correspond to 50 black tiles. However, this can be tedious. You might realize that adding 2 repeatedly is equivalent to multiplying by 2. Thus, if you know that in the fourth pattern, where there are 4 black tiles and 14 white tiles, you will need to add 46 more black tiles to get to 50 black tiles. Therefore, you add 2 white tiles 46 times for a total of an additional 92 white tiles, thus yielding 14 + 92 = 106 white tiles. If the pattern is exponential (1, 2, 4, 8, 16, 32, and so forth), you can use powers of 2 to represent multiplying by 2 repeatedly.

d. Based on this pattern, explain how you could find the number of white tiles needed for any number (n) of black tiles. Show or explain your work.

One way to answer this question is to use a table and treat the black tiles column as the "input" and the white tiles column as the corresponding "output." Thus, what "rule" can you generate that can take the "input" value and change it to the corresponding "output" value that would work for every pair in the table? You may need to try different combinations of operations to find the rule. In this example, the rule is: number of white tiles = $2n + 6$, where n = number of black tiles. You should check the rule to make sure that it works for any and all corresponding pairs in the table.

Another way to find the rule is to look at the pattern and notice that each black tile added corresponds with 2 white tiles added. But there are always 6 white tiles that make up the left and right vertical edges of each figure. Thus, $2n$ represents the repeated addition of 2 white tiles for every black tile. The + 6 represents the number of white tiles that stays the same in each figure, no matter how long the figure gets.

This equation (number of white tiles = $2n + 6$) is described as "direct" because it allows one to know the number of white tiles if given the number of black tiles without needing to find all the previous numbers of white tiles.

Name

PROBLEMS

1. The first four figures in a pattern are shown below:

Figure 1   Figure 2   Figure 3   Figure 4

a. Describe any recursive patterns in the figures shown above, assuming an ongoing pattern.

b. Based on the pattern, how can you find the number of squares needed for the 50th figure?

c. Based on the pattern, explain how you could find the number of squares needed for any figure number called n. Write a direct equation.

2. Find a direct equation for each input-output table below.

a.

IN (X)	OUT (Y)
1	3
2	6
3	9
4	12
5	15

c.

IN (X)	OUT (Y)
1	1
2	4
3	9
4	16
5	25

b.

IN (X)	OUT (Y)
1	2
2	5
3	8
4	11
5	14

d.

IN (X)	OUT (Y)
1	3
2	9
3	27
4	81
5	243

WRITING HEURISTICS, OR RULES OF THUMB

Heuristics are rules of thumb people follow in order to make judgments quickly and efficiently.

For each design challenge per class, keep a list on chart paper of research results and other suggestions to consider when making design decisions to meet criteria and constraints. The list should be revisited and added to or refined when students reflect on and discuss the results of their research. Encourage students to use the list when they brainstorm wand choose a design. The list should help them find a design that successfully meets the criteria and constraints.

EXAMPLE

Rules of Thumb for Design Challenge 1: Malaria Meltdown!
- Aluminum foil performed the least well as an insulator from heat (4 to 5 layers were needed to keep temperature below 30°C for 1 minute), but it is the thinnest material tested.

- You only need 2 layers either of bubble wrap or cardboard to keep the temperature below 30°F for 1 minute.

Other suggestions to consider when designing:
- Cardboard is sturdy and rugged, but gets soggy when wet.

AMAZON MISSION INTRODUCTION

OBJECTIVE: Students will read and understand the story line for the three design challenges in *Amazon Mission*.

> **CLASS**

ASK THE CLASS:

- What comes to your mind when I say the word *Amazon*?
  Possible Answer(s): Most students would probably say the online store Amazon.com. A few might know that it's the name for one of the world's largest rain forests. Explain that Jeff Bezos, the founder of Amazon.com, named his store *Amazon* because he envisioned that his store would become something big, just like the rain forest.

  You might also want to share that the Amazon region was given its name when Spanish explorer Francisco de Orellana, the first European to travel the whole length of the Amazon River, encountered a tribe of fierce warrior women and called them "amazonas," after a Greek myth about female warriors.

- Where is the Amazon rain forest located?
  Possible Answer(s): It's located in South America, including portions of Colombia, Ecuador, Peru, Bolivia, Brazil, and Venezuela. Tell students that they will be learning about a tribe living in the Amazon rain forest, and they will be helping its people solve some problems using engineering and math.

AMAZON MISSION INTRODUCTION

Deep in the heart of the tropical rain forests of Brazil and Venezuela lives a group of indigenous people known as the Yanomami (Yah-no-mah-mee). From the time they first inhabited the earth, the Yanomami lived in harmony with the environment, isolated from the rest of the world. They thrived in the lush rain forest, abundant with natural resources and wildlife—that is, until recently. In just the last few decades, outsiders have entered the Yanomami territory. This contact now threatens their very existence, and the Yanomami need your help! Imagine that you are joining the community service team at your school on a mission to the Amazon rain forest. You will investigate, and hopefully solve, the problems of the Yanomami. The future of the Yanomami people depends on you and your classmates!

- **INTRODUCING THE ENGINEERING DESIGN PROCESS (EDP)**
  What is engineering? What does an engineer do?

- **DESIGN CHALLENGE 1: MALARIA MELTDOWN!**
  Design a medicine carrier that can safely transport malaria medicine while keeping it cool in a tropical climate.

- **DESIGN CHALLENGE 2: MERCURY RISING!**
  Design a water-filtration system to remove mercury from a river.

- **DESIGN CHALLENGE 3: OUTBREAK!**
  Design a virus intervention plan to contain the spread of the flu.

AMAZON MISSION INTRODUCTION *(CONTINUED)*

OBJECTIVE: Students will read more background information about the Yanomami people.

1. [**CLASS**] Read aloud or ask students to read the next page aloud.

   ASK THE CLASS:

   - How have the people of the Yanomami tribe lived for nearly 20,000 years?

     Possible Answer(s): They live by hunting, gathering, fishing, gardening, and making necessary items such as baskets, hammocks, and bows and arrows. When not working, the adults engage in storytelling and spiritual activities.

   - How did encounters with outsiders affect the lives of the Yanomami tribe?

     Possible Answer(s): Gold miners spoiled the land of the Yanomami by cutting down trees, polluting the water and air with chemicals, and scaring off game (animals hunted by the Yanomami). Outsiders also introduced life-threatening diseases and killed Yanomami who opposed their activities. Thus, the Yanomami have become an endangered people.

2. [**CLASS**] Share with the class that the colorful patterns on the inside cover were inspired by traditional Yanomami designs.

AMAZON MISSION INTRODUCTION (CONTINUED)

YANOMAMI WAY OF LIFE

The community service team has done some initial research on the Yanomami people. They've found that there are approximately 20,000 Yanomami people currently living in Amazonia. The Yanomami live together in villages that may have as few as 40, and as many as 400, inhabitants. There are about 250 independent villages in the Amazon rain forest. In each village, the Yanomami people live together in a single, circular structure called a *shabono*. Within the shabono, families cluster together around their handmade hammocks and hearth fires. With no walls separating the families, privacy is rare in a Yanomami village. The center of the shabono consists of a large open space where the children play and the adults engage in celebrations.

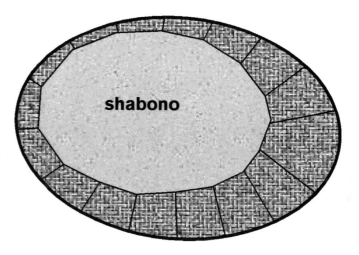

The Yanomami's simple way of life offers a glimpse into the way all humankind lived nearly 20,000 years ago. Their days consist of hunting, gathering, fishing, gardening, and making necessary items, including hammocks, baskets, and bows and arrows. In contrast to the industrialized world's digital age of music, television, and video, the Yanomami pass the time with storytelling and spiritual activities.

THREATS TO YANOMAMI EXISTENCE

In the 1960s, gold was discovered in the Yanomami territory. At that time, gold miners did not have the technology to effectively mine in a tropical rain forest. However, advancements in mining and travel technologies in the late 1900s eventually gave gold miners the means to enter and mine in Amazonia. With no government regulations to stop outsiders from entering the Yanomami land, gold-mining activity in the Amazon rain forest grew quickly. Miners in Amazonia had little concern for the Yanomami villagers or their land while on their search for gold. They cut down trees to make way for mining pits, polluted the water and air with chemicals, and scared off game with loud drilling. They introduced life-threatening diseases to the Yanomami, and even shot and killed some of the Yanomami villagers who got in their way.

Indeed, if life for the Yanomami continues in this way, it is estimated that the Yanomami people will become extinct in the next decade. You and your community service team must do what you can to save them!

Teacher Page

INTRODUCING THE ENGINEERING DESIGN PROCESS (EDP)

OBJECTIVE: Students will identify and order the steps of the engineering design process (EDP).

ESTIMATED TIME: 20 minutes

MATERIALS

- 1 set of EDP cards per pair or team of three to four students

BEFORE YOU TEACH

- Make sets of EDP cards by copying the EDP card templates (pages 28–29) back to back onto cardstock, cutting them out, shuffling them, and tying each set with a rubber band.
- Organize students into pairs or teams of three to four students.

PROCEDURES

1. To get a sense for what students know and think about engineering, ask: "What is an engineer? What does an engineer do?" Students can brainstorm and write their ideas in the space on page 30. If students are struggling to respond to these questions, ask them to list some things that are made by people—for example, houses, roads, cars, televisions, and phones. Explain that engineers have a part in the design and construction of all these things and many more.

2. Explain that all engineers use the engineering design process to help them solve problems in an organized way. Explain that students will use this design process to solve problems in the upcoming activities.

3. Distribute one set of EDP cards to each team. Instruct teams to distribute the cards evenly among themselves and take turns reading aloud each step's description. Their task is to correctly order the steps. Set the time limit to 3 minutes. When debriefing the activity, post each team's steps on the board and compare the lists. Where do teams agree and disagree? Where there is disagreement, ask the teams to explain their rationale for their particular orders. When revealing the "correct" order on pages 148–149, emphasize that the EDP is meant to be a set of guidelines to solve engineering design challenges, but engineers may not always follow all the steps in the same order all the time.

4. Ask the following questions to help students think more about these steps:

    - Why is the step "Communicate" part of the design process? How is it an important step?

      Possible Answer(s): It's important for engineers to communicate their design to other people so they can receive critical feedback and suggestions to improve the design.

    - What do you think happens after the last step, "Redesign"?

      Possible Answer(s): The engineer may go back to an earlier step—which could be as early in the process as "Identify the Problem"—depending on how well the prototype meets specifications. Once the design has gone through several cycles of the design process, it may then be produced on the full-scale level and constructed for real-world use.

5. The EDP matching exercise on page 31 gives students an opportunity to identify the EDP step used in a specific instance of a design challenge.

6. Wrap up the lesson by explaining that students will use these steps to solve three engineering challenges in *Amazon Mission* while learning and reinforcing their math skills and understanding. Point out the octagon in the right-hand corner of each page that shows where students are in the EDP. The descriptions of the EDP steps are also on pages 148–149.

DEFINE the problem. What is the problem? What do I want to do? What have others already done? Decide upon a set of specifications (also called "criteria") that your solution should have.

Conduct **RESEARCH** on what can be done to solve the problem. What are the possible solutions? Use the Internet, go to the library, conduct investigations, and talk to experts to explore possible solutions.

BRAINSTORM ideas and be creative! Think about possible solutions in both two and three dimensions. Let your imagination run wild. Talk with your teacher and fellow classmates.

CHOOSE the best solution that meets all the requirements. Any diagrams or sketches will be helpful for later EDP steps. Make a list of all the materials the project will need.

Use your diagrams and list of materials as a guide to **BUILD** a model or prototype of your solution.

TEST and evaluate your prototype. How well does it work? Does it satisfy the engineering criteria?

COMMUNICATE with your fellow peers about your prototype. Why did you choose this design? Does it work as intended? If not, what could be fixed? What were the trade-offs in your design?

Based on information gathered in the testing and communication steps, **REDESIGN** your prototype. Keep in mind what you learned from others in the communication step. Improvements can always be made!

RESEARCH	**DEFINE**
CHOOSE	**BRAINSTORM**
TEST	**BUILD**
REDESIGN	**COMMUNICATE**

INTRODUCING THE ENGINEERING DESIGN PROCESS (EDP)

1. What is engineering? What does an engineer do? Brainstorm and list some of your ideas in the space below.

2. Your team will be given some cards, each naming and describing a step in the engineering design process (EDP). Engineers use the EDP to solve design challenges, just like you will as you go through *Amazon Mission*. Your task is to put these steps in a logical order, from Step 1 to Step 8. Be prepared to explain your reasoning for the order you choose.

STEP 1: _____

STEP 2: _____

STEP 3: _____

STEP 4: _____

STEP 5: _____

STEP 6: _____

STEP 7: _____

STEP 8: _____

3. Imagine that you are part of a team that builds sails and uses them in boat races. Match each sentence to the appropriate step in the engineering design process.

SENTENCE	ENGINEERING DESIGN PROCESS STEP
a. You talk with other sailors to find out how their sails are made.	
b. Your team spends a Saturday making a new sail.	
c. After you win the race, you explain the design of the sail to your competitors.	
d. A race is coming up and your boat needs a new sail. The team decides that the sail must be waterproof, affordable, and strong enough to handle powerful winds.	
e. After your team meets to discuss the different designs, your team decides on one design. You find a marina that sells sail material that is strong, waterproof, and cheap.	
f. The sail works pretty well, but when strong gusts of wind blow, the seams rip. Your team resews the seams using stronger stitching.	
g. One week before the race, your team tests the new sail.	
h. Each person on your team sketches a sail design.	

Design Challenge 1
Malaria Meltdown!

INTRODUCTION

OBJECTIVE: Students will read and understand the problem presented for the first design challenge.

| CLASS | Together, read the introduction on the next page.

ASK THE CLASS:

• Given what you know about rain forests, why might it be difficult to keep the medicine between 10˚C and 30˚C?

Possible Answer(s): The climate in a rain forest is tropical, which means that the temperature is high—usually over 32˚C. So the challenge is to find a way to keep medicine cool in a hot climate.

INTERESTING INFO

There are five different climate zones:

1. **TROPICAL** climate zones have annual and monthly temperature averages above 20°C.

2. **SUBTROPICAL** climate zones have an average temperature range of 10°C to 20°C and four to eleven months with temperature averages above 20°C.

3. **TEMPERATE** climate zones have an average temperature range of 10°C to 20°C for four to twelve months.

4. **COLD** climate zones have an average temperature range of 10°C to 20°C for one to four months and the rest cooler.

5. **POLAR** climate zones have an average temperature below 10°C for all twelve months.

Design Challenge 1

Malaria Meltdown!

INTRODUCTION

One of the Yanomami villages has already contacted the community service team at your school, requesting your help. More than half of the villagers are sick with malaria, suffering from severe fevers, chills, and fatigue. Some of the sickest villagers are on the verge of death. The only thing that can save them now is antimalarial medication, but the village has run out! The Yanomami people have asked that you bring a new supply of medicine to the village, and now desperately await your arrival!

Malaria is spread through the bite of the anopheles mosquito. Abandoned mining pits, which are filled with still, warm water, are a breeding ground for these mosquitoes. It seems impossible to stop the spread of the disease. For now, the Yanomami people must rely on antimalarial medicine to treat their symptoms and cure them of the disease. Scientists have just developed a new drug that is 98% effective in curing malaria. The Yanomami need a supply of this new medicine immediately. The drug is highly sensitive. It must be kept between 15°C and 30°C at all times. If the medicine's temperature drops below 15°C or goes above 30°C, it becomes permanently ineffective. You need to transport the medicine to the village while controlling its temperature carefully.

1. DEFINE THE PROBLEM: MALARIA MELTDOWN!

OBJECTIVE: Students will read and understand the criteria and constraints of the design challenge.

CLASS Read about the design challenge on the next page so students understand the criteria and the constraints.

DEFINE

1. DEFINE THE PROBLEM: MALARIA MELTDOWN!

In order to get the medicine to the village, you and your classmates will first fly to Manaus, Brazil. From there, you will take a helicopter to a clearing that is approximately 13 km from the Yanomami village. The medicine will remain safe during the flights, stored in a large temperature-controlled refrigerator. However, once you've reached the helicopter landing, you must carry the medicine the rest of the way to the village by foot. The hike to the village will take 2 hours. The Amazon rain forest has an average temperature of 37°C this time of year. In the heat of the rain forest, medical officials think that their current medicine carrier will not be able to keep the medicine below 30°C for very long. They fear the medicine will spoil before it reaches the village. The hospital needs your engineering team's help to design a new medicine carrier that can keep the medicine between 15°C and 30°C for the entire hike!

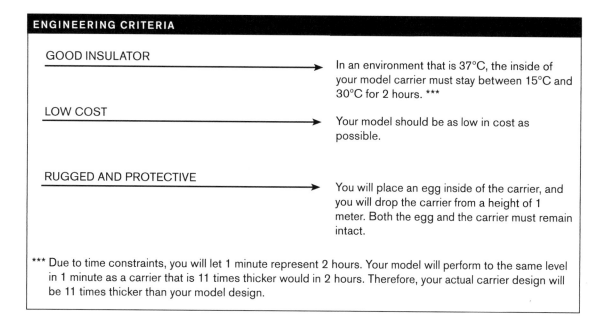

ENGINEERING CRITERIA

GOOD INSULATOR → In an environment that is 37°C, the inside of your model carrier must stay between 15°C and 30°C for 2 hours. ***

LOW COST → Your model should be as low in cost as possible.

RUGGED AND PROTECTIVE → You will place an egg inside of the carrier, and you will drop the carrier from a height of 1 meter. Both the egg and the carrier must remain intact.

*** Due to time constraints, you will let 1 minute represent 2 hours. Your model will perform to the same level in 1 minute as a carrier that is 11 times thicker would in 2 hours. Therefore, your actual carrier design will be 11 times thicker than your model design.

ENGINEERING CONSTRAINTS

You are limited to the following materials for your carrier design:

- corrugated cardboard
- foam board
- bubble wrap
- aluminum foil
- heavy-duty tape

2. RESEARCH THE PROBLEM: MALARIA MELTDOWN!
RESEARCH PHASE 1: ANALYZE THE CURRENT MEDICINE CARRIER

OBJECTIVES

Students will:
- interpret a line graph
- connect the research phase to the design challenge

> **CLASS** | Read the next page together.

ASK THE CLASS:

- Why is the laboratory room set at 37°C?
  Possible Answer(s): To simulate the average temperature of the Amazon rain forest

- How can this experiment's data help someone who's designing a medicine carrier for the Amazon?
  Possible Answer(s): The graph shows how quickly the temperature within a container increases over time when the container is exposed to an environmental temperature of 37°C. The data can help us determine whether or not the container can successfully transport the medicine without spoiling it.

TIP

Make sure students understand that the temperature of the thermometer starts at 15°C because the medicine was cooled to this temperature. The temperature will increase over time because the room temperature of the lab is 37°C.

---

2. RESEARCH THE PROBLEM: MALARIA MELTDOWN!
RESEARCH PHASE 1: ANALYZE THE CURRENT MEDICINE CARRIER

A team of engineers has already done some initial testing on the current medicine carrier. The team cooled the medicine to the minimum temperature of 15°C and then placed it inside the carrier. The carrier was then placed in a 37°C laboratory room. The team checked the temperature of the medicine every 10 minutes for 2 hours. The results of their test are provided in Table 1.1 and Graph 1.1 below.

Table 1.1: Temperature of Medicine Over Time in Current Container (in 37°C Lab Room)

TIME (MIN)	0	10	20	30	40	50	60 (1 hr)	70	80	90	100	110	120 (2 hr)
TEMP (°C)	15.0	20.2	24.2	27.2	29.4	31.2	32.5	33.5	34.3	34.8	35.3	35.6	35.9

Graph 1.1: Temperature of Medicine Over Time in Current Container (in 37°C Lab Room)

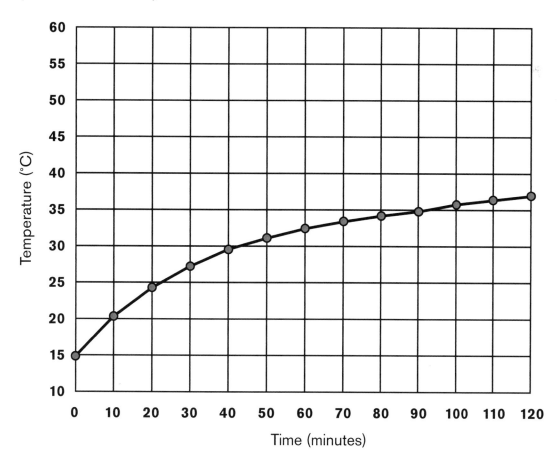

Teacher Page

2. RESEARCH THE PROBLEM: MALARIA MELTDOWN!
RESEARCH PHASE 1: ANALYZE THE CURRENT MEDICINE CARRIER
(CONTINUED)

OBJECTIVES

Students will:

- calculate and interpret the slope of a line
- describe the relationship between two variables
- graph a compound inequality
- connect the research phase to the design challenge

1.　TEAMS　Assign teams to use the table and graph on page 37 to answer questions 1–5 in 7 to 8 minutes. Note the following as you circulate around the room:

- Are students able to calculate the slope? Many students might only be familiar with calculating slope on a standard Cartesian plane that uses the abstract symbols of x and y. Do they recognize how "Time" corresponds to x and "Temperature" corresponds to y and use ΔTemperature/ΔTime to calculate the slope?

- Did students realize that the temperature indicated by the graph will eventually approach the room temperature of 37°C and not go above that?

- Were the questions too easy or too difficult?

- Did students have enough time to complete the questions?

- How well are the students working together?

- Were any student responses particularly insightful? Note those for whole-group sharing later.

2.　CLASS　Go over the questions and note the following points:
Review slope—both how to find it given two points on a graph and what it means. In this case, the slope represents an average rate of increase in temperature.

Teacher Page *(continued)*

ASK THE CLASS:

- How much do you predict that the temperature will rise from 120 minutes to 130 minutes?

  Possible Answer(s): Estimates should be less than 0.5 degrees, given the trend of the graph. The rate of increase appears to slow down over time, meaning that the amount of increase becomes less and less.

- Why does the temperature slow its rate of increase over time? Why doesn't the temperature keep rising at a steady rate?

  Possible Answer(s): Initially, the rate of increase is high because the difference between the room temperature and the temperature inside the container is so great. Heat travels to where there is less heat. However, as the temperature inside the container approaches the room temperature, the rate slows because the difference in temperature becomes less and less.

- Students may try to multiply 24 hours by a constant rate of deg./hour to get an answer for question 5 (for example, it rose 17.5°C in the first hour so it will rise 420°C in 24 hours). Why is this reasoning incorrect? Why would it not be appropriate to apply a constant rate of change to this situation?

  Possible Answer(s): The temperature inside the container will approach 37°C until it becomes 37°C, at which time the temperature will stop increasing and the rate of increase will be 0. It doesn't make sense for the temperature to increase at a steady rate because the temperature inside the container can't go above the room temperature.

3. **CLASS** First ask students to complete problem 6 on page 41 individually or in pairs. Review the answer together as a class. Ask students how they would represent the compound inequality on the graph on page 37. Instruct them to shade in the appropriate section of the graph with a colored pencil.

4. **TEAMS** Assign teams to discuss and complete questions 7 and 8 in 3 to 4 minutes.

5. **CLASS** Go over student responses to questions 7 and 8.

   ASK THE CLASS:

   - Does the medicine's temperature always have to be increasing, or can it also decrease?

     Possible Answer(s): As long as the temperature values are always within the acceptable range, you have a good carrier. If students have a graph that does decrease in temperature, they should have a rationale for how they might actually make this happen. For instance, the container might have a cooling device that self-activates when the temperature rises to a certain level and shuts off when it drops to another level.

2. RESEARCH THE PROBLEM: MALARIA MELTDOWN!
RESEARCH PHASE 1: ANALYZE THE CURRENT MEDICINE CARRIER
(CONTINUED)

Use Table 1.1 and Graph 1.1 to answer the questions below and on the next page.

1. Based on the graph, describe how the temperature is changing as time goes on.

    _____

    _____

    _____

2. Using a ruler, draw a straight line that goes through the data points at 0 and 10 minutes.

    What is the slope of this line? _____

    What does this slope tell you about what is happening to the medicine's temperature between 0 and 10 minutes?

    _____

    _____

    _____

3. Using a ruler, draw a straight line that goes through the data points at 110 and 120 minutes.

    What is the slope of this line? _____

    What does this slope tell you about what is happening to the medicine's temperature between 110 and 120 minutes?

    _____

    _____

    _____

4. Based on the graph and your answers to questions 2 and 3, describe how the rate of change in the temperature changes over time.

    _____

    _____

    _____

5. If the team of engineers left the medicine in the carrier in the 37˚C laboratory room for an entire day (24 hours), what do you think the medicine's temperature would be at the end of the day? _____˚C

    Explain your reasoning.

    _____

    _____

    _____

6. The temperature of the medicine must not drop below 15°C and must not rise above 30°C. Using the variable T to represent temperature, write these constraints in the form of a compound inequality on the line below.

    _____

    Using a colored pencil, graph this inequality directly on Graph 1.1, and shade the region that represents all acceptable temperatures for the medicine.

7. Compare the graph of the current medicine carrier to the shaded region of acceptable temperatures. How would you rate the effectiveness of the current carrier? Explain your answer.

    _____

    _____

    _____

8. Imagine a better medicine carrier that could keep the medicine between 15°C and 30°C for the entire 2-hour period. What might the graph of the temperature versus time look like for this carrier? Sketch one possible graph for this better medicine carrier directly on Graph 1.1, and label it "Better Carrier." There is more than one correct answer.

    How might you design a carrier that would match such a graph? Please describe your ideas in detail.

    _____

    _____

    _____

Teacher Page

2. RESEARCH THE PROBLEM: MALARIA MELTDOWN!
RESEARCH PHASE 2: INVESTIGATING DIFFERENT MATERIALS

OBJECTIVES

Students will:
- conduct a controlled experiment
- collect experimental data in a table

MATERIALS

For each team:
- 1 pack of colored pencils
- 1 digital thermometer
- 1 cup of ice
- 1 stopwatch
- 2 pieces of each material, about 15 cm × 15 cm each (corrugated cardboard, foam board, bubble wrap, aluminum foil)

1. **CLASS** Show students the different materials they can use in their designs. Ask them to predict which one will be the best material to keep heat out. Instruct them to read the instructions on the next page together.

2. **TEAMS** Give teams 10 to 15 minutes to test as many materials as they can. Emphasize the following:
   - Remind students that they will be taking all the data for each material in a 1-minute trial. At the end of each 1-minute trial, they should have collected temperature values at all five time intervals on the table.
   - Materials should not be folded or have overlapping pieces.
   - The meatiest part of the palms should be applying heat to the thermometer.

RESEARCH

2. RESEARCH THE PROBLEM: MALARIA MELTDOWN!
RESEARCH PHASE 2: INVESTIGATING DIFFERENT MATERIALS

Now that you have completed some research on the old design, you are ready to investigate some new materials for your design. You will use your body temperature to simulate the 37°C tropical climate of the Amazon.

Student jobs: Assign each member of your group to one of the four jobs listed below.

Heater: _____   **Timer:** _____
Cooler: _____   **Recorder:** _____

STEP 1: COOLER–Place the uncovered thermometer stem in ice. Cool thermometer to 10°C or colder.

STEP 2: HEATER–For the "bare hands" trial, place the thermometer stem between your palms. For trials with materials, place one layer of material on each palm and then place the thermometer stem between the pieces of material. Make sure that the materials completely cover the thermometer stem, including the tip. Also, do not fold the materials. Press palms together. When the thermometer reads 15°C, say "Start timing!" Keep your palms pressed firmly together for 1 full minute.

STEP 3: TIMER–When you hear the **Heater** say "Start timing!" immediately click "start" on your stopwatch. Shout "Time!" at every 10-second interval (when the watch reads 10, 20, 30, 40, 50, and 60 seconds).

STEP 4: HEATER–Watch the thermometer. Each time the **Timer** shouts "Time!" read the temperature aloud. The **Recorder** should record the values that the **Heater** reports in Table 1.2 below.

STEP 5: Repeat steps 1–4 with a different material.

Table 1.2: Temperature of Different Materials Over Time

MATERIAL	TEMPERATURE (°C) ON THERMOMETER						
	0 sec	10 sec	20 sec	30 sec	40 sec	50 sec	60 sec
1. bare hands	15°C						
2. corrugated cardboard	15°C						
3. foam board	15°C						
4. bubble wrap	15°C						
5. aluminum foil	15°C						

2. RESEARCH THE PROBLEM: MALARIA MELTDOWN!
RESEARCH PHASE 2: INVESTIGATING DIFFERENT MATERIALS
(CONTINUED)

OBJECTIVES

Students will:

- produce and analyze a graph that relates two variables
- determine when it's appropriate to use a line graph to represent data
- distinguish between independent and dependent variables
- connect this research phase to the design challenge

MATERIALS

For the class:

- overhead projector
- sheet of chart paper for the Rules of Thumb list (see page 21)

For each team:

- 1 copy of the graph on transparency
- 1 pack of thin transparency markers (at least four different colors)

1.　　**CLASS**　Instruct teams to read page 46. If the class needs practice graphing or scaling axes, you can use the activities on pages 5–20 as review or reinforcement. Provide the Rubric for Graphs on page 152 so students can assess their work.

    ASK THE CLASS:

    - What kind of graph should you make to represent your data? Why a line graph?

      Possible Answer(s): Because the data is continuous—temperature changes over time. Both time and temperature are measured using values in the real number set, as opposed to discrete values such as whole numbers that don't contain the parts in between whole numbers.

    - How should you label the axes? Which is the independent variable (x-axis)?

      Possible Answer(s): Time, because students controlled that variable

    - Which is the dependent variable (y-axis)?

      Possible Answer(s): Temperature, because it varies according to time

    - How do you represent four sets of data on a single graph?

      Possible Answer(s): Use a different color for each line.

2.   **TEAMS**   Teams that finish early may draw their graphs on transparencies to share their results with the class.

3.   **CLASS**   **ASK THE CLASS:**

   - Which material seems to be best at keeping heat out? How can you tell?

   - Show teams' graphs on the overhead projector. Put the transparencies on top of one another to compare. Do everyone's data agree? What might account for any differences in data results?

     Possible Answer(s): Differences occur due to human error in experimentation. Some factors are difficult to control. For example, the temperature of someone's hands might vary from person to person.

   - How well do you think more than 1 layer of material would perform to keep heat out? Which combinations of materials might work well together? After giving students some time to share their ideas, tell them that they will test different combinations of materials during the next research phase.

4.   **CLASS**   Introduce heuristics, or rules of thumb (see page 21). Ask students what they learned from this activity and Research Phase 2 that they can add to the Rules of Thumb list.

TIP
You might want to assign a color to each material so that all graphs can be read with the same key.

---

2. RESEARCH THE PROBLEM: MALARIA MELTDOWN!

RESEARCH PHASE 2: INVESTIGATING DIFFERENT MATERIALS

1. Graph your data from Table 1.2 on the grid below. Remember to label the axes, color in the key, and give the graph a title. Use the rubric provided by your teacher to assess your work. Then use your graph to answer the questions that follow.

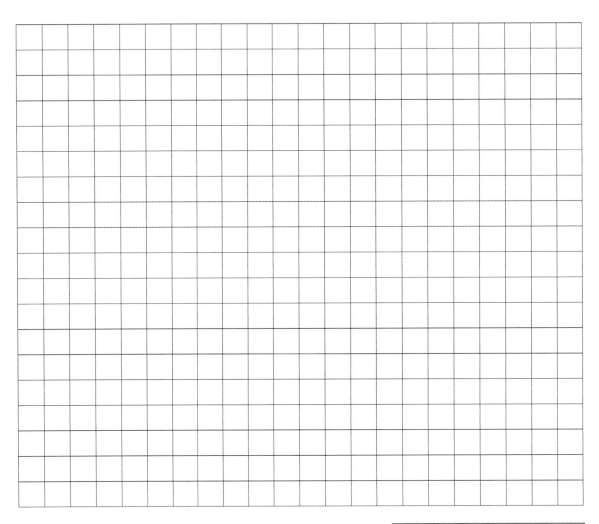

KEY

□ = bare hands

□ = corrugated cardboard

□ = foam board

□ = bubble wrap

□ = aluminum foil

2. Which of the materials, if any, would be able to keep the medicine from spoiling?

_____

_____

_____

3. Which material worked the best? _____

    Why do you think this material worked better than the others?

_____

_____

_____

4. Do you think there would be any drawbacks or disadvantages to using this material? Explain.

_____

_____

_____

2. RESEARCH THE PROBLEM: MALARIA MELTDOWN!
RESEARCH PHASE 3: COMBINING DIFFERENT MATERIALS

OBJECTIVES

Students will:
- conduct a controlled experiment
- collect experimental data in a table
- list permutations of up to five layers of two different kinds of materials
- produce a graph to present the data collected from testing combinations of materials
- connect this research phase to the design challenge

MATERIALS

For the class:
- overhead projector
- the Rules of Thumb list

For each team:
- 1 digital thermometer
- 1 cup of ice
- 1 stopwatch
- enough materials to make various permutations of layers for testing
- 1 copy of a blank graph on transparency
- 1 pack of thin transparency markers (at least 4 different colors)

1. **CLASS** Explain to students that they will experiment with different permutations of materials today. Instruct students to read page 50 together. Define *permutation*, if necessary.

   ASK THE CLASS:

   Based on your research results in Research Phase 2, which two materials would you choose to combine and why?

2. **TEAMS** Teams should choose two materials, write out four ways to combine the materials, make a prediction about which would perform the best, and conduct the experiment following the same procedures as in Research Phase 2. Teams that finish early may do additional permutations. Give teams 20 minutes to do this experiment.

3. **TEAMS** Instruct teams to display their results as a graph and to answer question 5. Provide the Rubric for Graphs on page 152 so students can assess their work. Give each team a transparency and markers so they can transfer their graph onto the transparency. Teams should prepare to give a short presentation of their graph and results.

4. **CLASS** Give each team 1–2 minutes to present their findings to the class.
   ASK THE CLASS: How can you use what you learned from the research phase to help you design the medicine carrier?
   Possible Answer(s): We can use the research results to find the combination of materials that can keep the temperature between 15°C and 30°C.

5. **CLASS** Add students' findings and suggestions from question 5 to the Rules of Thumb list.

ASSESSMENT
Use the Rubric for Graphs on page 152 to grade each team's graph.

OPTIONAL CCSS ENHANCEMENT
To address additional aspects of the Common Core State Standards, emphasize the manipulation of symbols related to permutations of materials in the medicine carrier.

---

2. RESEARCH THE PROBLEM: MALARIA MELTDOWN!
RESEARCH PHASE 3: COMBINING DIFFERENT MATERIALS

Now that you have investigated how well individual materials keep out the heat, you will take a look at how different materials work in permutation. You will test _____ and _____.

1. Come up with four different ways that you could combine these two materials. Each permutation can have anywhere from 2 to 5 total layers of material. For each permutation, identify the innermost and outermost layers.

	Permutation 1	Permutation 2	Permutation 3	Permutation 4
INNERMOST LAYER ↓ OUTERMOST LAYER				

2. Of the permutations you identified above, which do you think will be the best insulator? Explain your reasoning.

_____

_____

_____

3. Using the same steps that you used to conduct the experiment in Research Phase 2, test all four permutations you identified above. You will need two sets of materials for each permutation to make a "sandwich." Record your results in the table below.

Table 1.3: Temperature versus Time for Different Permutations of Materials

Material Permutations	Temperature (°C) on Thermometer						
	0 sec	10 sec	20 sec	30 sec	40 sec	50 sec	60 sec
1.	15°C						
2.	15°C						
3.	15°C						
4.	15°C						

4. You will be presenting your results to the class. Decide as a group how to best display the results of your experiment. Then use the grid below to display your data. Use the rubric provided by your teacher to assess your work.

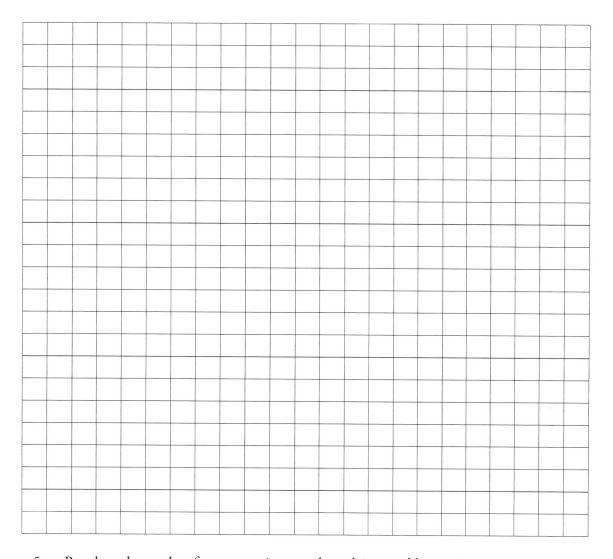

5. Based on the results of your experiment, what advice would you give to your classmates on designing their medicine carriers regarding the materials you tested?

_____

_____

_____

3. BRAINSTORM POSSIBLE SOLUTIONS: MALARIA MELTDOWN!

OBJECTIVES

Students will:
- review the criteria and constraints of the design challenge
- individually brainstorm a medicine-carrier design

1. **CLASS** Remind students of the engineering criteria and constraints on page 54. Each design can include up to five layers of materials. The cost for each square meter of material is listed in Table 1.4. Note the costs of the materials, and remind students that one of the criteria is to make a low-cost container. Discuss any other factors that might be important to consider. Is the carrier water-resistant? Easy to carry? Lightweight? Not too large?

2. **INDIVIDUALS** Instruct students to individually come up with a medicine-carrier design and draw the design on page 55. The carrier can be any shape. It does not have to be a rectangular prism. The design should include the choice of materials, number of layers, placement of materials, and shape of carrier. Review and use the Rules of Thumb list to help students make design decisions.

INTERESTING INFO

A vacuum flask, or thermos, is made by combining two containers. An outer container encloses an inner container. A vacuum exists between the inner and outer containers. This helps prevent rapid temperature changes for the liquid inside the inner container. Originally, vacuum flasks were made with glass. However, as technology advanced, they were made with metals.

INTERESTING INFO

Visit these sites for information on dimensioning drawings.

General dimensioning:
http://engineering.dartmouth.edu/mshop/pdf/introdr.pdf

Dimensioning circles and radius with AutoCAD:
www.youtube.com/watch?v=5fh3x6ZpUdM

3. BRAINSTORM POSSIBLE SOLUTIONS: MALARIA MELTDOWN!

You've done some great research and are ready to think about some possible medicine-carrier designs. As you brainstorm possible solutions, keep these design criteria and constraints in mind.

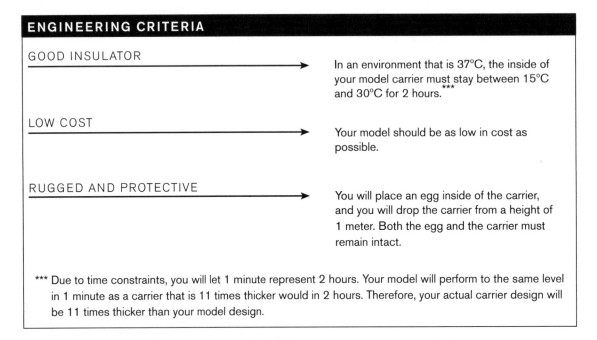

ENGINEERING CRITERIA

GOOD INSULATOR → In an environment that is 37°C, the inside of your model carrier must stay between 15°C and 30°C for 2 hours.***

LOW COST → Your model should be as low in cost as possible.

RUGGED AND PROTECTIVE → You will place an egg inside of the carrier, and you will drop the carrier from a height of 1 meter. Both the egg and the carrier must remain intact.

*** Due to time constraints, you will let 1 minute represent 2 hours. Your model will perform to the same level in 1 minute as a carrier that is 11 times thicker would in 2 hours. Therefore, your actual carrier design will be 11 times thicker than your model design.

ENGINEERING CONSTRAINTS

You may combine the five available materials in whatever manner you choose. The cost of 1 square meter (m²) of each material is listed in Table 1.4 below.

Table 1.4: Cost per Square Meter (m²) of Materials for Medicine Carrier

MATERIAL	COST FOR 1 SQUARE METER (m²)
corrugated cardboard	$0.99
foam board	$2.25
bubble wrap	$2.85
aluminum foil	$0.50

© Museum of Science (Boston), Wong, Brizuela

INDIVIDUAL DESIGN

Working on your own, use the results of your research and the cost information provided in Table 1.4 to imagine one possible medicine-carrier design. Sketch your idea in the space below. Label all materials and the dimensions (length, width, and height) of your carrier.

4. CHOOSE THE BEST SOLUTION: MALARIA MELTDOWN!

OBJECTIVES

Students will:

- share their individual brainstorm designs and decide on a team design
- draw nets of their medicine carriers
- find the surface areas of their nets to calculate the cost of materials for their medicine carriers

MATERIALS

For each team:

- net of a rectangular prism (optional)

CLASSROOM MANAGEMENT TIPS

- If students need more structure to share their individual medicine-carrier designs from the Brainstorm step, go over a list of things for each person to share (e.g., the materials chosen, the total cost, the order of layers, the number of layers); a set time for each person to share (about 1 minute); and behavior expectations for teammates during sharing (e.g., no interruptions, no comments until everyone has shared, active listening).
- Discuss with students how to come to an agreement on a team design. Team members can take turns discussing pros and cons of each design, identify commonalities in designs, compromise on areas of disagreement, and either vote or try to reach consensus.

ASSESSMENT

Introduce students to the Rubric for Engineering Drawings on page 155 by showing students examples of student work on pages 160–163, and using the rubric to grade each drawing. You can first assess one drawing with the whole class using the rubric and then have students work in pairs to assess the other drawings. Debrief as a whole class. Students can use the rubric to self-assess their own drawings of the medicine-carrier design.

INTERESTING INFO

Being able to sketch is an important ability to have in engineering. You will often need to make diagrams and keep track of measurements in projects. Here are some tips on sketching:

a. Draw using your shoulder rather than your wrist. Drawing with the shoulder allows you to make continuous lines in one straight movement. With less wrist movement, there will be fewer jagged edges and bumps.

b. Use a pad of layout paper. Layout paper has very light gridlines.

c. Use a pencil.

d. Draw a proportion scale in your sketch.

Teacher Page

BEFORE YOU TEACH (OPTIONAL)

Copy the Appendix (net of a rectangular prism) onto cardstock, one per team.

1. **TEAMS** Instruct teams to follow the instructions on page 58 to share their individual medicine-carrier designs and work as a team to use the best ideas to come up with a team design. Once each team agrees on a design, they may proceed to sketch their medicine-carrier design. Remind students to review the Rules of Thumb list to help them make design decisions that meet the criteria and constraints.

2. **CLASS** If needed, review what "nets" are and how to use them to find the area of a 3-D shape.

3. **CLASS** As an exercise, give each team the net of a rectangular prism that they can fold, unfold, and measure to find the surface area.

4. **TEAMS** Instruct teams to draw a net of their carrier design. Students should label all the lengths of their net.

5. **TEAMS** Students should find the surface areas of their nets and show their work in problem 4. Finally, they should calculate the cost of making their carrier by completing Table 1.5. Although it is a fairly straightforward calculation exercise, you might want to ask students to come up with the steps for calculating the cost of the materials they will use before they see Table 1.5. This is a multistep problem, although once the steps are established, it's a simple matter of making sure that the steps are correctly followed and the calculations are correctly done.

INTERESTING INFO

In order to create three-dimensional shapes, math nets are often used. Math nets are two-dimensional shapes on paper that, when folded correctly, provide a three-dimensional shape. Check out the following link for more on nets of a cube, a tetrahedron, an octahedron, an icosahedron, a dodecahedron, a pyramid, a cube octahedron, a buckyball, and a sphere: http://gwydir.demon.co.uk/jo/solid/.

OPTIONAL CCSS ENHANCEMENT

To address additional aspects of the Common Core State Standards, direct students to write out the equation for the total cost of the carrier.

CHOOSE

4. CHOOSE THE BEST SOLUTION: MALARIA MELTDOWN!

TEAM DESIGN

1. Have each member of your team share his or her ideas with the group. For each idea, think about the following:

    • What do you like the most about the design?

    • Do you think the design will meet all of the engineering criteria and constraints?

2. As a group, decide on one "best" solution. Draw a three-dimensional sketch of the medicine-carrier design in the space below. Then label the materials and the dimensions of the carrier (length, width, height, and so forth). Draw a cross section of the carrier to show the order of materials you are layering. Use the rubric provided by your teacher to check the quality and completeness of your drawing. Practice using the rubric on the sample drawings provided by your teacher.

3. Look at the design you drew on page 58. Imagine what this carrier would look like if it were completely opened up and unfolded onto a single, flat surface. This is called the net of your three-dimensional design. Draw this net as best you can in the space below. Label all side lengths of your net. Use the rubric provided by your teacher to check the quality and completeness of your drawing. Practice using the rubric on the sample drawings provided by your teacher.

EXAMPLE

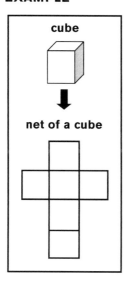

4. Using the net you drew in problem 3, calculate the approximate surface area of your medicine-carrier design. This surface area will help you find the approximate cost of each layer of material in your carrier.

SHOW YOUR WORK

5. Using the table below, calculate the cost of each material and the total cost of your design.

Table 1.5: Calculating Cost

MATERIAL	NUMBER OF LAYERS NEEDED	AREA OF EACH LAYER (YOUR ANSWER TO PROBLEM 4)	TOTAL AREA FOR ALL LAYERS OF THIS MATERIAL (m^2)	COST PER SQUARE METER (m^2)	TOTAL COST FOR EACH MATERIAL ($)
corrugated cardboard				$0.99	
foam board				$2.25	
bubble wrap				$2.85	
aluminum foil				$0.50	

Total cost: _____

5. BUILD A PROTOTYPE/MODEL: MALARIA MELTDOWN!

OBJECTIVE: Students will build prototypes of their medicine carriers.

1. **CLASS** Together, read page 62, which explains that students will build a model of the medicine carrier and note any changes to their design in the space provided. Provide the Rubric for Prototype/Model on page 157 so students can assess their work.

2. **TEAMS** Each team should send one member to gather the appropriate materials for their design.

ASSESSMENT

Use the Rubric for Prototype/Model on page 157 to assess completeness and craftsmanship of students' models.

INTERESTING INFO

During World War II, there was a huge demand for blood on the battlefield to help save the lives of wounded soldiers. However, blood needs to be stored at a certain temperature (4.4°C to 10°C). This was especially a problem in the tropical temperatures of the Pacific. Dr. Henry Blake designed the "Frigichest" to store blood at this temperature range. He created his Frigichest by using an insulated cardboard container wrapped in aluminum foil. He also added a carrying strap to the box so that medics could easily move the Frigichest to and from different sites.

---

5. BUILD A PROTOTYPE/MODEL: MALARIA MELTDOWN!

Following your design from Step 4, gather all necessary materials and construct a model of your medicine carrier. As you build your model, draw or use words to describe any changes to your original design in the space below. Use the rubric provided by your teacher to assess your work.

6. TEST YOUR SOLUTION: MALARIA MELTDOWN!

OBJECTIVE: Students will follow testing procedures to test their medicine-carrier prototypes.

MATERIALS

For each team:
- 1 egg (raw)
- 1 resealable sandwich bag
- 1 digital thermometer
- 1 cup of ice
- 1 stopwatch
- 1 meterstick

CLASS Have each team follow the directions on the next page to test their design. Provide the Rubric for Test, Communicate, and Redesign Steps on page 158 so students can assess their work.

> **ASSESSMENT**
> Use the Rubric for Test, Communicate, and Redesign Steps on page 158 to assess how well students followed testing procedures.

6. TEST YOUR SOLUTION: MALARIA MELTDOWN!

You may use the rubric provided by your teacher to assess your work on the next few pages.

INSULATOR TEST

Student jobs: Assign each member of your team to one of the four jobs listed below.

Heater: _____   **Timer:** _____

Cooler: _____   **Recorder:** _____

STEP 1: **COOLER**–Place uncovered thermometer stem in ice. Cool thermometer to 10°C or colder.

STEP 2: **HEATER**–Get extra pieces of the materials you used for your design. Layer the materials as you did when you constructed your carrier. Place one stack of materials in each palm, and then place the thermometer stem between the stacks. Make sure that the materials cover the entire thermometer stem, including the tip. Press palms together. When the thermometer reads 15°C, say "Start timing!" Keep your palms pressed firmly together for 1 full minute.

STEP 3: **TIMER**–When you hear the **Heater** say "Start timing!" immediately click "start" on your stopwatch. Shout "Time!" after exactly 1 minute (60 seconds).

STEP 4: **HEATER**–Watch the thermometer. When the **Timer** shouts "Time!" read the temperature aloud.

STEP 5: **RECORDER**–Record the temperature after 1 minute: _____ °C

Did the thermometer stay between 15°C and 30°C for 1 minute?

☐ yes   ☐ no

RUGGED/PROTECTIVE TEST

Place a raw egg in a sandwich bag inside your carrier. Seal the bag shut. Measure a height of 1 meter from the ground, and drop your medicine carrier from this height.

Did the egg remain intact (no cracks)?   ☐ yes   ☐ no

Did the carrier remain intact?   ☐ yes   ☐ no

LOW-COST TEST

In order to keep the medicine at an acceptable temperature for 2 hours in the 37°C Amazon rain forest, the actual carrier would have to be 11 times thicker than your design that works for 1 minute. This means you will need 11 times more materials for the actual carrier than for your model. How much would this actual carrier cost? _____

Show your work below. Compare this cost to other teams' solutions.

7. COMMUNICATE YOUR SOLUTION: MALARIA MELTDOWN!

OBJECTIVE: Students will answer questions to reflect on their designs and present their designs to the class.

1. **TEAMS** After teams test their medicine carriers, instruct them to complete questions 1–5. Students should be prepared to share their answers with the class.

2. **TEAMS** Give each team 2 to 3 minutes to share their medicine-carrier design and test results with the class. Encourage students to ask questions or offer suggestions to improve each design.

7. COMMUNICATE YOUR SOLUTION: MALARIA MELTDOWN!

1. Do you think that your medicine-carrier design was successful? Did it meet all of the criteria and constraints? Explain your answer.

    _____

    _____

    _____

2. Specifically, what are some strengths or advantages of your design? Explain.

    _____

    _____

    _____

3. What are some drawbacks or disadvantages of your design? Explain.

    _____

    _____

    _____

4. If you could use any materials on Earth, what materials would you use to make your medicine carrier?

    _____

    Explain why you would choose these materials. _____

    _____

    _____

5. If you were going to sell your medicine carrier, how much would you sell it for? How would you market your carrier? What advertising or slogans would you use to make people want to buy it?

    _____

    _____

    _____

Be prepared to present your answers to questions 1–5 to the class.

© Museum of Science (Boston), Wong, Brizuela

8. REDESIGN AS NEEDED: MALARIA MELTDOWN!

OBJECTIVE: Students will answer questions to consider how they can redesign their medicine carriers.

1.   **TEAMS** Teams should answer questions 1 and 2 to improve their designs based on their classmates' suggestions and their own reflection.

2.   **CLASS** Wrap up the activity.

   ASK THE CLASS:

   • What math skills do you use in this activity?

     Possible Answer(s):

     We constructed line graphs to represent data, calculated slope, analyzed the graph to obtain useful information to help us design the carrier, drew nets of three-dimensional shapes, calculated surface area and cost of materials, and used measurement tools to construct the medicine carrier.

   • Why is a line graph useful for representing data?

     Possible Answer(s):

     A graph can easily show the relationship (if any) between two sets of data. It can also show trends—whether upwards or downwards. The steepness of the line can show how quickly or slowly the data is changing.

   • How do you determine which variable is dependent and which is independent? Give examples.

     Possible Answer(s):

     The independent variable is the one that you control, such as the number of layers or the time intervals. The dependent variable is the data collected that corresponds with the independent variable, such as the temperature.

   • What does the slope tell us about the graph? Give an example.

     Possible Answer(s):

     The slope can tell us the average rate of change between two points on the graph. For instance, if the graph relates time and temperature, the slope can tell us how much the temperature is changing per time unit.

8. REDESIGN AS NEEDED: MALARIA MELTDOWN!

1. Based on the tests of your medicine-carrier design, what changes could you make to improve it?

    _____

    _____

    _____

    Explain how these changes would improve your design.

    _____

    _____

    _____

2. Identify one thing that you learned from another group's medicine-carrier design that you can use to improve your design.

    _____

    _____

    _____

    Explain how this will improve your design.

    _____

    _____

    _____

INDIVIDUAL SELF-ASSESSMENT RUBRIC: MALARIA MELTDOWN!

OBJECTIVE: Students will use a rubric to individually assess their involvement and work in this design challenge.

ASSESSMENT

Assign this reflection exercise as homework. You can write your comments on the lines below the self-assessment and/or use this in conjunction with the Student Participation Rubric on page 159.

INDIVIDUAL SELF-ASSESSMENT RUBRIC: MALARIA MELTDOWN!

Use this rubric to reflect on how well you met behavior and work expectations during this activity. Check the box next to each expectation that you successfully met.

LEVEL 1	LEVEL 2	LEVEL 3	LEVEL 4	BONUS POINTS
Beginning to meet expectations	Meets some expectations	Meets expectations	Exceeds expectations	
☐ I was willing to work in a group setting.	☐ I met all of the Level 1 requirements.	☐ I met all of the Level 2 requirements.	☐ I met all of the Level 3 requirements.	☐ I helped resolve conflicts on my team.
☐ I was respectful and friendly to my teammates.	☐ I recorded the most essential comments from other group members.	☐ I made sure that my team was on track and doing the tasks for each activity.	☐ I helped my teammates understand the things that they did not understand.	☐ I responded well to criticism.
☐ I listened to my teammates and let them fully voice their opinions.	☐ I read all instructions.	☐ I listened to what my teammates had to say and asked for their opinions throughout the activity.	☐ I was always focused and on task: I didn't need to be reminded to do things; I knew what to do next.	☐ I encouraged everyone on my team to participate.
☐ I made sure we had the materials we needed and knew the tasks that needed to be done.	☐ I wrote down everything that was required for the activity.	☐ I actively gave feedback (by speaking and/or writing) to my team and other teams.	☐ I was able to explain to the class what we learned and did in the activity.	☐ I encouraged my team to persevere when my teammates faced difficulties and wanted to give up.
	☐ I listened to instructions in class and was able to stay on track.	☐ I completed all the assigned homework.		☐ I took advice and recommendations from the teacher about improving team performance and used feedback in team activities.
	☐ I asked questions when I didn't understand something.	☐ I was able to work on my own when the teacher couldn't help me right away.		☐ I worked with my team outside of the classroom to ensure that we could work well in the classroom.
		☐ I completed all the specified tasks for the activity.		

Approximate your level based on the number of checked boxes: _____ Bonus points: _____

Teacher comments: _____

© Museum of Science (Boston), Wong, Brizuela

TEAM EVALUATION: MALARIA MELTDOWN!

OBJECTIVE: Students will evaluate and discuss how well they worked in teams.

ASSESSMENT

1. **INDIVIDUALS** Assign this reflection exercise as homework or during quiet classroom time.

2. **TEAMS** Instruct students to share their team evaluation reflections with one another and discuss how they can improve their teamwork during the next activity.

3. **CLASS** Point out any good examples of teamwork and areas to improve during the next activity.

TEAM EVALUATION: MALARIA MELTDOWN!

How well did your team work together to complete the design challenge? Reflect on your teamwork experience by completing this evaluation and sharing your thoughts with your team. Celebrate your successes and discuss how you can improve your teamwork during the next design challenge.

RATE YOUR TEAMWORK. On a scale of 0–3, how well did your team do? 3 is excellent, 0 is very poor. Explain how you came up with that rating.

LIST THINGS THAT WORKED WELL. Example: We got to our tasks right away and stayed on track.

LIST THINGS THAT DID NOT WORK WELL. Example: We argued a lot and did not come to a decision that everyone could agree on.

HOW CAN YOU IMPROVE TEAMWORK? Make the action steps concrete. Example: We need to learn how to make decisions better. Therefore, I will listen and respond without raising my voice.

Teacher Page

Design Challenge 2
Mercury Rising!

INTRODUCTION

OBJECTIVE: Students will read and understand the problem presented for the second design challenge.

CLASS   Tell the class: "In the last activity, we designed a medicine carrier that would protect the malaria medicine from the heat during the trip to the Yanomami tribe. Now imagine that we're approaching the area where they live." Together, read the introduction on the following page. You might want to briefly explain that even though each fish may only ingest a tiny bit of mercury, a person can accumulate a poisonous amount of mercury by consuming many fish over a period of time.

ASK THE CLASS:

How would you advise the Yanomami to respond to this problem?
> **Possible answer(s):**
> Ask the gold miners to use another chemical that doesn't contain mercury; find a way to remove the mercury from the water.

INTERESTING INFO
Exposure to mercury can lead to mercury poisoning. Short-term, limited exposure to mercury can cause bleeding gums, vomiting, and stomach pain. Mercury also builds up in the body over time, because the body cannot easily get rid of mercury. As mercury builds up in the body, the results can be fatal—brain damage, liver damage, kidney damage, and even death may occur.

Design Challenge 2

Mercury Rising!
INTRODUCTION

As you approach the Yanomami village to deliver the antimalarial medicine, you notice a man mining for gold along the banks of the Amazon River. The miner does not seem to be disturbing anyone. However, as you approach the Yanomami village, you hear the following pieces of conversation:

"If the gold miners poison the water and kill the fish, where will we drink and what will we eat?"
—Jose Siripino, Yanomami

"We want progress without destruction. We want to study, to learn new ways of cultivating the land and living from its fruits. The Yanomami do not want to live from dealing with money, with gold."
—Davi Kopenawa, Yanomami

The Yanomami people have been expecting you, and are grateful for the medicine. After getting the medicine into the proper hands, they include you in their conversation about the gold miners. They explain that the gold miners are contaminating the freshwater of the Amazon. The miners use the chemical mercury to help them collect grains of gold along the river. However, not all the mercury is taken up by the gold, and some of it is released into the river. Fish, a main food source for the Yanomami, are being poisoned with the mercury. This poison passes on to the villagers through the food chain and leads to a variety of illnesses.

1. DEFINE THE PROBLEM: MERCURY RISING!

OBJECTIVE: Students will read and understand the criteria and constraints of the design challenge.

 CLASS Together, read the following page and make sure that students understand the criteria and the constraints.

INTERESTING INFO

There are several types of filters that can filter out mercury from water. One particular type is the activated carbon (AC) filter. There are two subgroups of the AC filter—the granular activated carbon and solid-block activated. The granular activated carbon filter is packed with granules of activated carbon. The solid-block activated carbon filter is made of activated carbon particles compressed into a dense material. When water travels through these filters, mercury and other contaminants are trapped.

OPTIONAL CCSS ENHANCEMENT

To address additional aspects of the Common Core State Standards, direct students to express the water flow criteria as inequalities.

---

1. DEFINE THE PROBLEM: MERCURY RISING!

The Yanomami people presented their argument to the local miners. Reluctantly, the miners agreed to use a portion of their profits to fund a water-filtration system that will filter out at least 75% of the mercury in the freshwater near the mining operation. However, they will not provide any money for the filtration system until they receive a design plan that meets some requirements.

Your task is to design a small-scale water-filtration system that can be scaled up to an actual system. Your smaller model only has to filter 250 mL of water. For the actual system, the miners will hire contractors to build a system that is several thousand times larger.

ENGINEERING CRITERIA

LOW COST	Your filter design should be as low in cost as possible.
FILTERING RATE	Your design must filter at least 540 liters of water per day.
FILTERING ABILITY	Your design should filter no more than 1 liter of water per minute (so that water is flowing slowly enough to ensure that 75% of the mercury is removed).

ENGINEERING CONSTRAINTS

You may use any or all of the following materials to design and build your filter:

- plastic bag
- plastic stirrers
- plastic straws
- small and large cups
- tape
- scissors

Teacher Page

2. RESEARCH THE PROBLEM: MERCURY RISING!
RESEARCH PHASE 1: MINIMIZING COST

OBJECTIVES

Students will:
- make a prediction about which size Mercatrons* will meet the minimum surface-area requirement and have the lowest cost
- calculate the surface area of a sphere using a formula
- solve a multistep problem

MATERIALS

For each team:
- a sheet of chart paper for the Rules of Thumb list (see page 21)

1. **CLASS** Read about Mercatrons on page 79. Point out that the 400 square centimeters refers to the surface area of the spheres. Introduce problem 1, in which teams are asked to make a prediction as to which size Mercatrons would result in the least cost.

2. **TEAMS** As teams work on problem 1, ask them to think about what they would need to know to figure out which size sphere would result in the least cost and still have enough total surface area.

3. **CLASS** Call on teams to share their predictions and reasoning in response to problem 1.

4. **TEAMS** Give teams time to think about what steps they would need to follow to solve this problem before moving on to the next page.

5. **CLASS** Ask teams to share their reasoning.

INTERESTING INFO

*Mercatrons are a fictional substance that is factually based. NUCON International, Inc. makes coal-based pellets called Mersorb® that can absorb mercury from gas, water, and liquid hydrocarbons.

Teacher Page *(continued)*

6. **CLASS** Together, read problem 2 and review how to use the formula to find the surface area of a sphere. You might want to ask, "Why do we need to know the surface area of the Mercatrons?" Guide students through working out the reasoning for how to figure out which size sphere would meet the minimum surface-area criteria and be the lowest in cost:

   a. Find the surface area of one Mercatron sphere of a particular size.

   b. Find how many spheres of that size are needed to achieve 400 square centimeters (cm^2).

   c. Calculate how many packs of spheres would need to be bought and their total costs.

   d. Compare the total cost needed to buy x number of spheres of each size to see which is the lowest in cost.

7. **PAIRS** Assign pairs to complete problems 2 and 3 in about 7 to 8 minutes. Pairs that finish early may check their answers with the other pair in their team. As you circulate, see whether students understand that they need to round up when determining the total packages needed. You can ask, "If you need 12 Mercatrons, how many packages do you need to buy?"

8. **CLASS** Go over answers to problems 2 and 3, or collect a sampling of activity sheets to assess. Also, add the result from problem 3 to the Rules of Thumb list (see page 21).

OPTIONAL CCSS ENHANCEMENTS

To address additional aspects of the Common Core State Standards, encourage students to translate verbal descriptions of mathematical relationships into equations using letters. Emphasize to students that one or more parts of an expression can be viewed as a single entity when students compute the cost of Mercatrons. Direct students to calculate the surface area of Mercatrons needed to filter 100% of the mercury. Guide students to calculate the volume of the Mercatron spheres (in addition to the surface area).

---

2. RESEARCH THE PROBLEM: MERCURY RISING!
RESEARCH PHASE 1: MINIMIZING COST

Scientists have discovered a material that absorbs mercury. This material has been coated over the surface of different sized spheres. These mercury-absorbing spheres are called *Mercatrons.* The costs per package of the different sized Mercatrons are listed in the table below.

Table 2.1: Cost of Mercatrons

MERCATRON DIAMETER (CM)	COST PER PACKAGE OF 10 MERCATRONS
0.4	$0.15
0.8	$0.30
1.2	$0.60
1.6	$1.20
2.0	$2.40
2.4	$4.80

The miners have agreed to pay for the removal of 75% of the mercury in a small region of the Amazon River near the mining operations. For your scaled-down model, scientists tell you that the contaminated water must have contact with a minimum of 400 square centimeters (cm^2) of the surface of the Mercatron before exiting the filter in order for at least 75% of the mercury to be dissolved.

1. Make a prediction about which size of Mercatron will give the miners the lowest cost for the filtration system: _____

    Explain your prediction.

    _____

    _____

    _____

2. Perform the necessary calculations to complete the table below. Show your work and round your answers to the nearest hundredth. To find the surface area of each Mercatron size, use the formula:

surface area of a sphere = 4 • 3.14 • (radius)^2
(3.14 is an approximation of π)

Table 2.2: Total Cost for Different Sized Mercatrons

Diameter of Mercatron (cm)	0.4	0.8	1.2	1.6	2.0	2.4
Radius of Mercatron (cm)						
Surface area of Mercatron (cm^2)						
Total number of Mercatrons needed for a total surface area of 400 cm^2						
Total number of packages needed (10 Mercatrons per package)						
Cost per package of 10 Mercatrons	$0.15	$0.30	$0.60	$1.20	$2.40	$4.80
Total cost						

WORKSPACE

3. Which Mercatron size has the lowest total cost? _____
   Compare this answer to your prediction in problem 1.

Teacher Page

2. RESEARCH THE PROBLEM: MERCURY RISING!
RESEARCH PHASE 2: MINIMUM AND MAXIMUM FILTRATION RATES

OBJECTIVES

Students will:

- convert measurement units (within the same system)
- use proportional reasoning
- write a compound inequality statement

MATERIALS

For the class:

- the Rules of Thumb list

1. **CLASS** Introduce problems 1 and 2, which involve calculating the time (in seconds) that it should take 250 mL of water to flow through the filter according to the criteria. You may also assign this problem for homework.

TIP

Students may need an activity to remind them how to convert units. See pages 14–16 for lesson plan and exercises.

---

2. **PAIRS** Assign pairs to work together to solve problems 1 and 2. If students are having trouble getting started, ask them how they would write out in fraction form "540 liters of water per day" and "seconds per 250 mL." Ask them how they would convert "day" to "seconds" and "liter" to "mL." Students should note that at some point, they have to "flip" the volume/time rate so that it becomes time/volume. Also, they may need to be reminded that 1 liter = 1000 mL.

3. **CLASS** Go over problems 1 and 2. You may want to select two pairs of students to explain their solutions to the class. Tell the class to follow the reasoning of the explanation to see if it makes sense, and ask students to agree or disagree with the explanation. Allow the class to direct any questions to the pair explaining their answer. Also, ask the class if anyone used a different method to solve the problems.

If students are wondering why the minimum rate seems to be a higher number (40 sec/250 mL) than the maximum rate (15 sec/250 mL), you can explain that the minimum rate refers to the slowest speed that the water should flow through the filter. Thus, a longer time (such as 40 seconds) would correspond with a slower speed. Meanwhile, the maximum rate refers to the fastest that the water can flow (and still allow for 75% mercury removal), and thus 15 seconds would correspond with the fastest possible speed.

Add the results from problem 3 to the Rules of Thumb list.

2. RESEARCH THE PROBLEM: MERCURY RISING!
RESEARCH PHASE 2: MINIMUM AND MAXIMUM FILTRATION RATES

You will test the model of your filter design with 250 mL of water. Therefore, it is important to figure out what is an acceptable amount of time to filter 250 mL, based on the engineering criteria.

1. **Minimum rate:** Your filter must filter at least 540 liters of water per day. Convert this flow rate into seconds per 250 mL.

SHOW YOUR WORK

2. **Maximum rate:** The water must flow at a rate of less than 1 liter per minute. Convert this flow rate into seconds per 250 mL.

SHOW YOUR WORK

3. Based on the flow rates you found in problems 1 and 2, write a compound inequality to represent the range of acceptable numbers of seconds to filter exactly 250 mL of water.

Teacher Page

2. RESEARCH THE PROBLEM: MERCURY RISING!
RESEARCH PHASE 3: INVESTIGATING FLOW RATE

OBJECTIVES

Students will:
- conduct a controlled experiment
- collect experimental data in a table

MATERIALS

For each team:
- 1 chopstick
- 1 large nail
- 1 pin
- 1 small nail
- 4 large (12-oz) foam cups
- measuring cup with 250-mL line
- ruler (metric)
- stopwatch

1. **CLASS**

   ASK THE CLASS:

   - Given a foam cup to serve as the filter, how can you control how quickly the water flows through the filter?

     Possible Answer(s): Two possible factors that one can vary are the size of the hole and the number of holes. Explain that students will investigate the relationship between the size of the hole and how quickly the water flows through the hole.

   - Read and demonstrate the experiment instructions on page 87. Emphasize that students should begin timing when the water first starts to exit the cup and flow into the measuring cup.

   - Which hole size do you predict will give a flow rate that is within the acceptable range (between 15 and 40 seconds)? What do you think will be the relationship between hole size and flow rate?

2. **TEAMS** Assign 20 minutes for teams to conduct the experiments.

Teacher Page *(continued)*

OBJECTIVES

Students will:
- produce and analyze a graph that relates two variables
- distinguish between independent and dependent variables
- connect the results of this research to the engineering design challenge
- determine when it is appropriate to use a line graph to represent data

1. **CLASS** Provide the Rubric for Graphs on page 152 so students can assess their work. If the class needs practice graphing or scaling the axes, use the activities on pages 5–13 as review or reinforcement.

   ASK THE CLASS:
   - How would you label the axes? Why?

     Possible Answer(s): The *x*-axis should be labeled "Diameter (mm)" because the diameter is the independent variable. By controlling that variable, we can then determine its effect on the time, which is the dependent variable. The *y*-axis should be labeled "Time (seconds)."

   - Time is typically the independent variable. Discuss why, in this case, the time measurements are actually the dependent variable.

     Possible Answer(s): Time is the dependent variable because it changes depending on the size of the hole.

   - Discuss whether a line graph would be appropriate and why. Is diameter considered continuous data?

     Possible Answer(s): It may be considered continuous data because measurements for diameter can be in the real number set, not just whole numbers. A scatter plot may also be appropriate.

2. **TEAMS** Instruct teams to graph their results and to answer questions 2–4.

Teacher Page *(continued)*

OBJECTIVES

Students will:

- brainstorm other factors that may affect flow rate
- design and conduct a controlled experiment
- collect experimental data in a table

1. **CLASS** After going over the answers to questions 2–4, invite students to add to the Rules of Thumb list and discuss question 5.

   ASK THE CLASS:

   - Which of these factors do you think will have the greatest effect on flow rate? Why?
   - How can you design an experiment that would test just one factor? How do you control for just one variable?

2. **TEAMS** Instruct teams to choose one factor that they want to test and complete problem 6, in which they design an experiment that tests that one factor. Each team needs the teacher's approval before conducting the experiment. Check to see that students are controlling for one variable and that their experiment design is clear and has step-by-step procedures.

3. **TEAMS** Once teams receive approval for their experiment design, they may conduct the experiment and collect data, graph the results, and answer question 9 on page 91.

ASSESSMENT

Use the Rubric for Experiment Design on pages 153–154 to assess the completeness and quality of the graph, test procedures, and data analysis.

INTERESTING INFO

The scientific method is a series of steps that scientists often follow to solve problems. These steps include the following:

1. Make observations.
2. Create a hypothesis from the observations.
3. Use the hypothesis to make predictions.
4. Test the predictions by making more observations or test by experimentation.
5. Analyze results and draw conclusions, if possible.

Teacher Page *(continued)*

OBJECTIVES

Students will:
- produce and analyze a graph that relates two variables
- connect the results of this research to the engineering design challenge

TEAMS Ask each team to summarize their findings for the class and make recommendations for controlling the flow rate based on the results from their experiment. Add recommendations to the Rules of Thumb list. Students should use the Rubric for Graphs on page 152 to assess their work. In addition, they can use the Rubric for Experiment Design on pages 153–154 to assess the completeness and quality of the graph, test procedures, and data analysis.

Challenge students to be critical of their data results and conclusions by asking questions such as the following:
- How you know if your results are valid?
- Do you believe that you will get the same results if you repeat your experiments?
- How can you improve your experiment procedures to make your findings more valid?

INTERESTING INFO

Engineers use a lot of math in their work. The basic mathematical concepts and algebraic equations allow engineers to formulate theories and perform calculations. Using statistics and probability, engineers can test their hypotheses by analyzing data where samples tested may contain variability or where experimental error is likely to occur. Engineers use data analysis, such as filtering and coding information, to describe, summarize, and compare the data with their initial hypotheses. Finally, engineers use modeling and simulations to predict the behavior and performance of their designs before they are actually built.

OPTIONAL CCSS ENHANCEMENT

To address additional aspects of the Common Core State Standards, show students how to represent the tabular data as a double number line diagram.

© Museum of Science (Boston), Wong, Brizuela

2. RESEARCH THE PROBLEM: MERCURY RISING!
RESEARCH PHASE 3: INVESTIGATING FLOW RATE

Now that you have figured out the range of acceptable flow rates, you can try to figure out what type of filter will give you an acceptable flow rate. You will investigate how the size of the filter's outlet relates to flow rate.

Student jobs: Assign each member of your group to one of the four jobs listed below.

Heater: _____   **Timer:** _____

Pourer: _____   **Measurer:** _____

STEP 1: MEASURER–Get a cup, and poke a hole in the center of the bottom of the cup using one of the four instruments listed in Table 2.3. Then, using a ruler, measure the diameter of the hole you created in millimeters and record it in Table 2.3.

STEP 2: HOLDER–Hold the cup over the measuring cup. The cup's hole should be facing directly downward into the measuring cup. Cover the hole with your finger.

STEP 3: POURER–Start pouring water into the cup until the cup is almost full. When the **Holder** releases his or her finger and water is coming out from the hole, say, "Start timing!" Keep pouring water into the cup to keep the water at the same level during the entire trial.

STEP 4: TIMER–When you hear the **Pourer** shout, "Start timing!" start the timer.

STEP 5: MEASURER–Watch the measuring cup. If the water is dripping out at a very slow rate, when the water level reaches 50 mL, shout, "Stop timing!" If the water is coming out at a fast rate, when the water level reaches exactly 250 mL, shout, "Stop timing!"

STEP 6: TIMER–When you hear the **Measurer** shout, "Stop Timing!" stop the timer. If the **Measurer** stopped the time when the water level reached 50 mL, multiply the time (in seconds) by 5 and record the time value in Table 2.3. If the **Measurer** stopped the time when the water level reached 250 mL, record the time (in seconds) in Table 2.3.

STEP 7: Repeat steps 1–6 with a new hole-poking instrument from Table 2.3.

Table 2.3: Flow Rates for Different Size Outlet Holes

Outlet hole was made by	Diameter of outlet hole (mm)	Time for 250 mL to flow through cup (seconds)
pin		
small nail		
large nail		
chopstick		

2. RESEARCH THE PROBLEM: MERCURY RISING!
RESEARCH PHASE 3: INVESTIGATING FLOW RATE

1. Graph your data from Table 2.3 on the grid below to show the relationship between the diameter of the outlet hole and the time it takes for 250 mL of water to flow through the cup. Remember to label the axes and give the graph a title. Use the rubric provided by your teacher to assess your work. Then use your graph to answer the questions that follow.

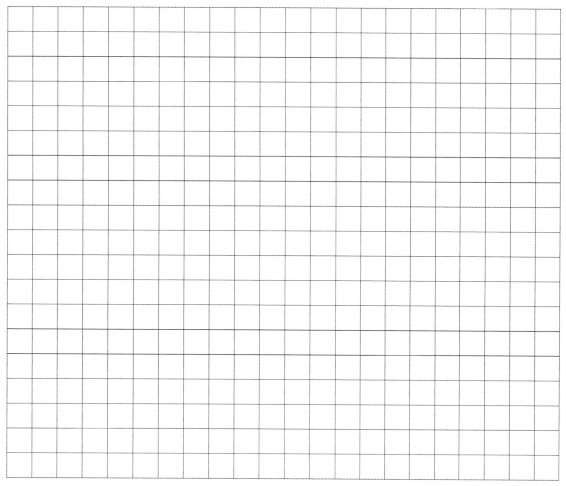

2. Describe the relationship between the diameter of the outlet hole and the time it takes for 250 mL of water to flow through the cup.

_____

_____

_____

3. Which diameter size(s) produce acceptable flow rates for the water filter? _____

4. Based on your graph, what size diameter would create a flow rate that is exactly in the center of the range of acceptable flow rates?

    _____

5. What factors, other than outlet size, might affect the flow rate of the water through the filter? List at least four ideas.

    _____

    _____

    _____

    _____

6. Consider the factors you listed in question 5. Put a star next to the factor you would like to test. As a group, create an experiment you could do to test this factor. Then, describe, in step-by-step detail, how you will go about testing this factor (use additional paper if needed).

    _____

    _____

    _____

7. After you have received approval from your teacher, conduct the experiment you designed above. Draw a table in the grid below, give it the appropriate labels, and record the results of your experiment in the table.

© Museum of Science (Boston), Wong, Brizuela

8. You will be presenting the results of your experiment to the class. Decide as a group how to best display your results. Then use the grid below to display your data. Remember to label the axes and give the graph a title. Use the rubric for graphs to assess your work.

9. Based on the results of your experiment, what advice would you give to your classmates about designing their water filters?

_____

_____

_____

Use the rubric for experiment design provided by your teacher to assess your work for problems 6–9.

3. BRAINSTORM POSSIBLE SOLUTIONS: MERCURY RISING!

OBJECTIVES
Students will:
- review the design criteria and constraints
- individually brainstorm and sketch a water-filter design

CLASS Explain that students have completed the research phase and are moving on to the brainstorm phase. Read the next page together and review the engineering criteria and available materials. Ask students to think about other filters that they have seen. How are they shaped? How do they work?

INDIVIDUALS Instruct students to individually brainstorm for 8 minutes. Students should review the research results and Rules of Thumb list, sketch a water-filter design, and label all the parts.

INTERESTING INFO
States and cities have different organizations to provide water to residents. For example, the Massachusetts Water Resources Authority (MWRA) provides water for much of the Massachusetts area. The water comes from two naturally filled reservoirs—the Quabbin Reservoir and the Wachusett Reservoir. The surrounding lands are protected so that the areas around both reservoirs are kept clean. When the water from the reservoirs reach the water treatment plant, the water is disinfected with ozone gas bubbles. Chloramines are then added to protect the water from contamination as it travels through the pipes. The pH level of the water is also adjusted so that it is not too acidic or basic. Finally, fluoride is added to the water for healthy teeth.

BRAINSTORM

3. BRAINSTORM POSSIBLE SOLUTIONS: MERCURY RISING!

You've done some great research and are ready to think about some possible water-filter designs. As you brainstorm possible solutions, keep the design criteria and constraints in mind.

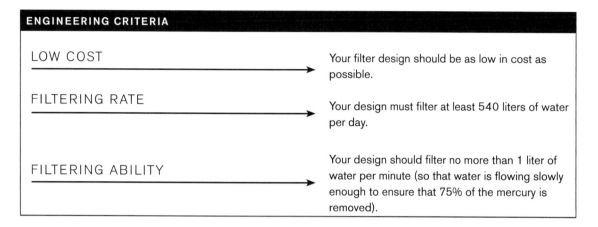

ENGINEERING CRITERIA	
LOW COST	Your filter design should be as low in cost as possible.
FILTERING RATE	Your design must filter at least 540 liters of water per day.
FILTERING ABILITY	Your design should filter no more than 1 liter of water per minute (so that water is flowing slowly enough to ensure that 75% of the mercury is removed).

ENGINEERING CONSTRAINTS

You may use any or all of the following materials to design and build your filter:

- plastic bag

- plastic stirrers

- plastic straws

- small and large cups

- tape

- scissors

Table 2.1: Cost of Mercatrons

MERCATRON DIAMETER (CM)	COST PER PACKAGE OF 10 MERCATRONS
0.4	$0.15
0.8	$0.30
1.2	$0.60
1.6	$1.20
2.0	$2.40
2.4	$4.80

INDIVIDUAL DESIGN

On your own, think about what your filter might look like and how it might work. What will be the overall shape of the filter? Where will the water go in? Where will the water come out? How big will the opening be? Where will the Mercatrons be located? Describe and draw your ideas in the space below. Label the inlet, the outlet, and the Mercatrons in your drawing.

© Museum of Science (Boston), Wong, Brizuela

4. CHOOSE THE BEST SOLUTION: MERCURY RISING!

OBJECTIVE: Students will share their individual brainstorm designs and decide on a team design.

TEAMS Give teams the remainder of the class time to share their ideas and come up with a team design that they want to build and test. Remind them to review the Rules of Thumb list when making design decisions. As students are working, circulate to make sure that their designs are complete—meaning that they are labeled with dimensions and locations of the Mercatrons, the inlet, the outlet, and so forth. Provide the Rubric for Engineering Drawings on page 156 so students can assess their work.

CLASSROOM MANAGEMENT TIPS

- If students need more structure to share their individual filter designs from the Brainstorm step, go over a list of things for each person to share (e.g., the number of holes, the size of holes, the position of holes, any other design elements, and the rationale for the design); a set time for each person to share (about 1 minute); and behavior expectations for teammates during sharing (e.g., no interruptions, no comments until everyone has shared, active listening).
- Discuss with students how to come to an agreement on a team design. Team members can take turns discussing pros and cons of each design, identify commonalities in designs, compromise on areas of disagreement, and either vote or try to reach consensus.

ASSESSMENT

Use the Rubric for Engineering Drawings on page 156 to assess students' work.

CHOOSE

4. CHOOSE THE BEST SOLUTION: MERCURY RISING!

TEAM DESIGN

Discuss your ideas as a team, and then decide upon one "best" solution. Draw the water filter design in the space below. Then label the materials to be used and the functions of various parts of the design. Use the rubric provided by your teacher to check the quality and completeness of your drawing.

© Museum of Science (Boston), Wong, Brizuela

5. BUILD A PROTOTYPE/MODEL: MERCURY RISING!

OBJECTIVE: Students will build their teams' water-filter designs.

MATERIALS

For each team (according to their design):
- plastic bag
- plastic stirrers
- plastic straws
- small and large foam cups
- 1 roll packing tape
- 1–2 pairs of scissors
- 1 ruler
- 90 small plastic spheres, 1.2 cm in diameter, to serve as "Mercatrons;" Mardi Gras beads work well

CLASS Go over the directions on the next page. As students build, remind them to write down any changes to their original design. Provide the Rubric for Prototype/Model on page 157 so students can assess their work.

ASSESSMENT

Use the Rubric for Prototype/Model on page 157 to assess completeness and craftsmanship of students' models.

---

5. BUILD A PROTOTYPE/MODEL: MERCURY RISING!

Following the design you chose in Step 4, gather all necessary materials and construct a model of your water filter. As you build your model, draw or use words to describe any changes to your original design. Use the rubric provided by your teacher to assess your work.

6. TEST YOUR SOLUTION: MERCURY RISING!

OBJECTIVE: Students will follow testing procedures to test their water-filter prototypes.

MATERIALS

For each team:

- 1 measuring cup with a 250-mL line
- 1 stopwatch
- water supply (about 500 mL—a standard bottle of water)

TEAMS During the testing phase, students will follow the instructions on the next page. Provide the Amazon Mission Rubric for Test, Communicate, and Redesign Steps on page 158 so students can assess their work. After completing the Test phase, students may use the remaining class time to discuss the questions on page 102.

ASSESSMENT

Use the Rubric for Test, Communicate, and Redesign Steps on page 158 to assess how well students followed testing procedures.

TEST

6. TEST YOUR SOLUTION: MERCURY RISING!

You may use the rubric provided by your teacher to assess your work on the next few pages.

FLOW-RATE TEST
Student jobs: Assign each member of your group to one of the four jobs listed below.

Heater: _____   Timer: _____

Pourer: _____   Measurer: _____

STEP 1: HOLDER–Place the measuring cup inside of the water basin. To begin, hold the water filter so that the opening where the water exits the filter is directed into the basin, but NOT into the measuring cup.

STEP 2: POURER–Pour water at a steady rate into the filter. Continue this for the entire test.

STEP 3: HOLDER–Watch the water exit your filter. Once the water seems to be flowing out of the filter at a steady rate, move the filter so that the water flowing out goes directly into the measuring cup and say, "Start timing!"

STEP 4: TIMER–When you hear the Holder shout, "Start timing!" start the timer.

STEP 5: MEASURER–Watch the measuring cup. When the water level reaches exactly 250 mL, shout, "Stop timing!"

STEP 6: TIMER–When you hear the Measurer shout, "Stop timing!" stop the timer.

How long did it take to filter 250 mL of water? _____

Is this time within the acceptable range?   ☐ yes   ☐ no

Would your filter be able to filter at least 540 liters of water per day?   ☐ yes   ☐ no

Does your filter design filter no more than 1 liter of water per minute?   ☐ yes   ☐ no

7. COMMUNICATE YOUR SOLUTION: MERCURY RISING!

OBJECTIVE: Students will answer questions to reflect on their designs and present their designs to the class.

1.   **TEAMS**   Teams should discuss and complete questions 1–4 as they prepare to present their design, test results, and reflections to the class.

2.   **CLASS**   Teams should give presentations of 2 to 3 minutes. Encourage the other teams to ask questions and give suggestions for improving their design.

7. COMMUNICATE YOUR SOLUTION: MERCURY RISING!

1. Do you think that your water-filter design was successful? Did it meet all of the criteria and constraints? Explain.

   _____

   _____

   _____

2. Specifically, what are some of the strengths or advantages of your filter design?

   _____

   _____

   _____

3. What are some drawbacks or disadvantages of your current design?

   _____

   _____

   _____

4. If you made a full-size version of your filter design, what materials would you use? Why would you choose these materials?

   _____

   _____

   _____

   Be prepared to present your answers to questions 1–4 to the class.

Teacher Page

8. REDESIGN AS NEEDED: MERCURY RISING!

OBJECTIVE: Students will answer questions to consider how they can redesign their water filters.

TEAMS Explain that designs can always be improved. Assign teams 7 to 8 minutes to answer questions 1 and 2.

ASK THE CLASS:

- What factors affect the flow rate of a water filter?
  Possible Answer(s): The factors include hole size, number of holes, and placement of holes.

- How did you investigate these factors?
  Possible Answer(s): We investigated the relationship between the size of holes and flow rate by making different size holes in different cups and timing how quickly water flows through the cup.

- How did you apply the results of your research to your water-filter design?
  Possible Answer(s): We used what we learned to determine how big the holes should be and how many to put in our filter to get the right flow rate.

- What math skills did you use in this activity?
  Possible Answer(s): We used inequalities to represent the flow-rate criteria and calculated the surface area of a sphere to determine the size of Mercatrons that would produce the lowest cost but still meet the surface-area criteria. We converted equivalent units, represented numerical data with graphs, analyzed the graphs for relevant results, and used measurement tools to build our design.

8. REDESIGN AS NEEDED: MERCURY RISING!

1. Based on the test of your water filter-design, what changes could you make to improve it?

    _____

    _____

    _____

    _____

    Explain how these changes would improve your filter design.

    _____

    _____

    _____

    _____

2. Identify one thing that you learned from another group's water-filter design that you could use to improve your filter design.

    _____

    _____

    _____

    Explain how this would improve your filter design.

    _____

    _____

    _____

    _____

INDIVIDUAL SELF-ASSESSMENT RUBRIC: MERCURY RISING!

OBJECTIVE: Students will use a rubric to individually assess their involvement and work on this design challenge.

ASSESSMENT

Assign this reflection exercise as homework. You can write your comments on the lines below the self-assessment and/or use this in conjunction with the Student Participation Rubric on page 159.

TEAM EVALUATION: MERCURY RISING!

OBJECTIVE: Students will evaluate and discuss how well they worked in teams.

ASSESSMENT

1. **INDIVIDUALS** Assign this reflection exercise as homework or during quiet classroom time.

2. **TEAMS** Instruct students to share their team evaluation reflections with one another and discuss how they can improve their teamwork during the next activity.

3. **CLASS** Point out any good examples of teamwork and areas to improve during the next activity.

STUDENT PAGE

INDIVIDUAL SELF-ASSESSMENT RUBRIC: MERCURY RISING!

Use this rubric to reflect on how well you met behavior and work expectations during this activity. Check the box next to each expectation that you successfully met.

LEVEL 1	LEVEL 2	LEVEL 3	LEVEL 4	BONUS POINTS
Beginning to meet expectations	Meets some expectations	Meets expectations	Exceeds expectations	
☐ I was willing to work in a group setting.	☐ I met all of the Level 1 requirements.	☐ I met all of the Level 2 requirements.	☐ I met all of the Level 3 requirements.	☐ I helped resolve conflicts on my team.
☐ I was respectful and friendly to my teammates.	☐ I recorded the most essential comments from other group members.	☐ I made sure that my team was on track and doing the tasks for each activity.	☐ I helped my teammates understand the things that they did not understand.	☐ I responded well to criticism.
☐ I listened to my teammates and let them fully voice their opinions.	☐ I read all instructions.	☐ I listened to what my teammates had to say and asked for their opinions throughout the activity.	☐ I was always focused and on task: I didn't need to be reminded to do things; I knew what to do next.	☐ I encouraged everyone on my team to participate.
☐ I made sure we had the materials we needed and knew the tasks that needed to be done.	☐ I wrote down everything that was required for the activity.	☐ I actively gave feedback (by speaking and/or writing) to my team and other teams.	☐ I was able to explain to the class what we learned and did in the activity.	☐ I encouraged my team to persevere when my teammates faced difficulties and wanted to give up.
	☐ I listened to instructions in class and was able to stay on track.	☐ I completed all the assigned homework.		☐ I took advice and recommendations from the teacher about improving team performance and used feedback in team activities.
	☐ I asked questions when I didn't understand something.	☐ I was able to work on my own when the teacher couldn't help me right away.		☐ I worked with my team outside of the classroom to ensure that we could work well in the classroom.
		☐ I completed all the specified tasks for the activity.		

Approximate your level based on the number of checked boxes: _____ Bonus points: _____

Teacher comments:

TEAM EVALUATION: MERCURY RISING!

How well did your team work together to complete the design challenge? Reflect on your teamwork experience by completing this evaluation and sharing your thoughts with your team. Celebrate your successes and discuss how you can improve your teamwork during the next design challenge.

RATE YOUR TEAMWORK. On a scale of 0–3 how well did your team do? 3 is excellent, 0 is very poor. Explain how you came up with that rating. Was it the same, better, or worse than the last activity?

LIST THINGS THAT WORKED WELL. Example: We got to our tasks right away and stayed on track.

LIST THINGS THAT DID NOT WORK WELL. Example: We argued a lot and did not come to a decision that everyone could agree on.

HOW CAN YOU IMPROVE TEAMWORK? Make the action steps concrete. Example: We need to learn how to make decisions better. Therefore, I will listen and respond without raising my voice.

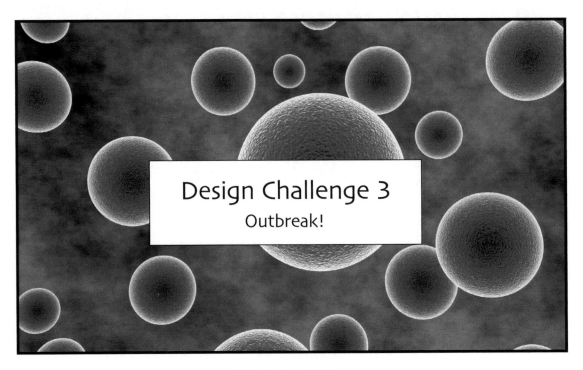

Design Challenge 3
Outbreak!

INTRODUCTION

OBJECTIVE: Students will read and understand the problem presented for the third design challenge.

CLASS Remind students that one of the effects of outsiders encountering the Yanomami is that the outsiders bring diseases to the region. Ask students, "What do you know about how a virus such as the common cold can spread?" Read the introduction to the design challenge on the next page.

INTERESTING INFO

Influenza causes about 20,000 deaths per year in the United States. However, death from influenza itself almost never happens. Usually, the influenza can aggravate other underlying problems, such as lung or heart disease. Because influenza invades the respiratory tract, bronchitis, sinusitis, and pneumonia can also occur.

The symptoms are not immediately evident when a person is infected. Instead, after a day or two, the symptoms will appear. Typical symptoms of influenza are a sore throat, dry cough, stuffed or runny nose, chills, fever, aching muscles and joints, headache, loss of appetite, nausea and vomiting, and fatigue. Normal recovery time is about a week.

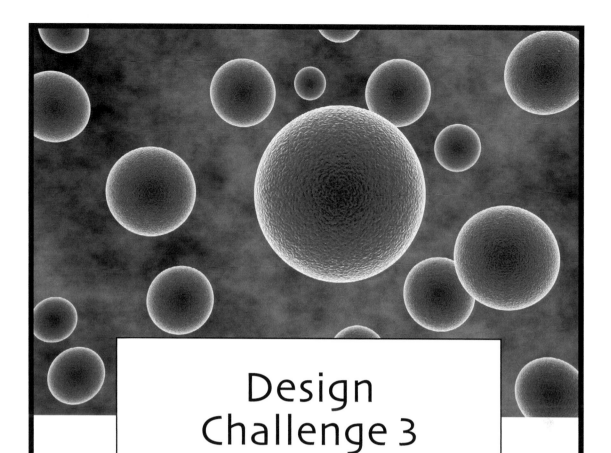

Design Challenge 3

Outbreak!

INTRODUCTION

The influenza virus (commonly known as the flu) was one of the illnesses that killed many Yanomami people in the 1980s. Influenza is particularly dangerous because it can spread so easily. When infected people cough or sneeze, they spray drops of the virus into the air. If other people breathe in these drops, or touch something that the drops have landed on (such as door handles or telephones), they are likely to become sick, too.

The influenza virus remains a threat to the Yanomami people because there are still outsiders near their villages, and the virus changes over time. Even if the villagers have been exposed to the flu in the past, they are not immune to new strains of the virus.

Teacher Page

1. DEFINE THE PROBLEM: OUTBREAK!

OBJECTIVE: Students will read and understand the criteria and constraints of the design challenge.

CLASS Read about the design challenge, criteria, and constraints on the next page.

ASK THE CLASS:
- The first criterion states that the number of villagers infected never goes above 25%. How many villagers are represented by that 25%?
  Possible Answer(s):
  10 villagers

- Given what you know about the different interventions, what combination do you think will make a good plan? Are doctors most important in stopping the spread of the virus, or will one of the preventive interventions be more effective?

INTERESTING INFO
Currently, there are no specific cures that exist for influenza. The best treatment is to increase consumption of fluids, such as water and juice, and get lots of bed rest. Certain drugs exist on the market that lessen the severity of the symptoms, but the primary efforts currently are in using vaccines to create immunity.

OPTIONAL CCSS ENHANCEMENT
To address additional aspects of the Common Core State Standards, direct students to calculate means for the infection rates of each of the virus intervention methods. As students conduct their applet simulations, guide a discussion about variability and/or the shape of data distributions for several trials with the same conditions.

1. DEFINE THE PROBLEM: OUTBREAK!

The Yanomami people need you to design a plan to protect them from a new strain of the flu virus that has been introduced to their village. Your plan needs to contain the virus for 30 days, but it also needs to be as low in cost as possible since the Yanomami people have little access to money.

ENGINEERING CRITERIA

VIRUS CONTAINED ———————————→ The percent of villagers infected does not go above 25% for 30 days.

LOW COST ———————————→ The plan is as low in cost as possible. You have a maximum budget of $10,000, but try to find a lower cost plan that works.

HIGH CHANCE OF SUCCESS ———————————→ The plan must meet the virus-containment criterion at least four out of five trials.

ENGINEERING CONSTRAINTS

- There are a total of 40 people in the Yanomami village that you will be helping. All of the villagers live together in a circular shabono that has a radius of about 16 meters.

- For our purposes, we will assume that without any protection, villagers have a 100% chance of catching the virus if they come into contact with a sick person.

- You may include any or all of the following interventions in your plan:

INTERVENTION	EFFECT OF INTERVENTION	COST OF INTERVENTION
doctor	can treat 1 person per day	$1,600 per doctor
air-filtration masks	reduces chance of getting virus by 50%	$750 for a village supply
antiviral hand gel	reduces chance of getting virus to 25%	$1,000 for a village supply
vaccinations	reduces starting chance of getting virus to 4%	$120 for 1 villager

Teacher Page

2. RESEARCH THE PROBLEM: OUTBREAK!
RESEARCH PHASE 1: HOW DOES A VIRUS SPREAD?

OBJECTIVES

Students will:
- participate in a simulation and collect simulation data
- identify and extend exponential patterns
- generalize and represent a pattern using symbols

MATERIALS
- 1 lump of clay or modeling compound

1. **CLASS** Together, read question 1. Discuss students' predictions and rationales.

2. **CLASS** Follow these directions to conduct the virus simulation:
   - The time (days) indicated in the table are considered "fast" days. The teacher announces "time" to fast-forward to the next time interval.
   - At time = 0, choose 1 person in the class to be sick, and give this person a large piece of clay. Everyone else in the class is healthy, but has a 100% chance of getting sick if a sick person gives them the virus.
   - At time = 1, the sick person will split the clay into two halves. He or she will keep one half and give the other half to someone in the class. This person is sick, so at time = 1, number sick = 2. Have students keep track of number sick.
   - Emphasize that students must accept the clay. They do not have the choice to refuse.
   - At time = 2, both sick people will split their clay in half and give one half away to infect someone new, so at time = 2, number sick = 4.
   - Continue this process until everyone in the class is sick. If you want to keep the simulation going one time interval longer, tell all sick students that their right hands represent a sick person and their left hands represent a healthy person. Their right hands can infect their left hands.

3. **CLASS** Ask the class to note any patterns in the growing number of sick people in the table. Ask, "How would the pattern continue?" Have students continue filling in the table according to the pattern they've noticed.

4.   **TEAMS**   Instruct teams to complete problems 3–5, in which they generalize the patterns they found and use their rules to make predictions. Students will plot their data using a colored pencil on the grid on page 122. You might want to discuss which is the independent variable (x) and the dependent variable (y).

In problem 4, many students may struggle to recognize and represent the pattern as powers of 2. Some students may attempt to represent the doubling pattern, which is a recursive equation that relies on knowing the number of sick people at time interval ($t - 1$) in order to find the number of sick people at time (t). While validating this pattern representation, encourage students to consider how they can find the number of sick people directly at any time interval. For instance, can they answer question 5 without having to extend the table to 20 days? For additional help and exercises on representing patterns with equations, use the activities on pages 17–20.

5.   **TEAMS**   Go over student answers to problems 3–5.

Amazon Mission

2. RESEARCH THE PROBLEM: OUTBREAK!
RESEARCH PHASE 1: HOW DOES A VIRUS SPREAD?

1. Imagine that one person in the classroom has the virus. Every time interval ("day"), the infected person can spread the virus to one other person. Anyone that becomes infected with the virus can then spread the virus to one new person every time interval. How much time do you think it will take for the entire class to be infected? Explain your prediction.

   _____

   _____

2. You will conduct an experiment as a class. Record the results of the experiment in Table 3.1 below.

Table 3.1: How Virus Spreads Through the Classroom

TIME (DAYS)										
NUMBER OF SICK PEOPLE										

3. Describe any patterns that you notice about the data. Try to find more than one pattern.

   _____

   _____

   _____

4. Represent the pattern(s) you found using pictures, symbols, or variables.

5. How many people do you think would be sick after 20 days? _____

   How did you determine this? _____

   _____

   _____

Teacher Page

2. RESEARCH THE PROBLEM: OUTBREAK!
RESEARCH PHASE 2: THERE'S A DOCTOR IN THE HOUSE

OBJECTIVES

Students will:

- participate in a simulation and collect simulation data
- identify and extend exponential patterns
- generalize and represent a pattern using symbols

MATERIALS

- 1 lump of clay or modeling compound

1. CLASS Discuss student predictions and rationales to question 1.

2. CLASS Follow these directions to conduct the simulation:

- Once again, these are "fast" days, meaning that they only need to last as long as it takes for the sick people to spread the virus during each round.
- At time = 0, choose 1 person to become sick and give that person a piece of clay.
- At time = 1, the sick person splits the clay and infects 1 more person.
- At time = 2, the 2 sick people each infect 1 new person. With 4 people sick, it seems like an epidemic might be starting, so a doctor is called; assign 1 student to be the doctor. The doctor will treat 1 person by taking his or her clay. Therefore, at the end of time = 2, number sick = 3.
- At time = 3, the 3 sick people each infect 1 new person. Then the doctor can again treat 1 person, so at the end of time = 3, number sick = 5.
- This will continue until all villagers become sick. As with Reseach Phase 1, you can continue the simulation by letting each arm of each student represent a person.
- *Note:* Treated students can get sick again.
- Students may ask why the doctor doesn't become sick. Explain that the doctor is vaccinated, and if he or she happens to get sick, the doctor can treat himself or herself.

Teacher Page *(continued)*

3. **TEAMS** Instruct teams to complete problems 3–5. Many students will have difficulty generalizing the pattern. Have them look for similarities between Table 3.2 and Table 3.1. See if students can observe that in Table 3.2, when time = 1, the number sick is 1 more than the number of sick in Table 3.1, time = 0. The number of sick in Table 3.2 is always 1 more than the number of sick in Table 3.1, shifted one time interval earlier. Ask students how they can use this observation, plus their rule from Research Phase 1, to generate a rule for the new pattern.

TIME(DAYS)	0	1	2	3	4	5	6	7	8	9	10	11
NUMBER OF SICK PEOPLE (NO DOCTOR)	1 +1	2 +1	4 +1	8 +1	16 +1	32 +1	64 +1	128 +1	256 +1	512 +1	1024 +1	2048
NUMBER OF SICK PEOPLE (WITH DOCTOR)	1	2	3	5	9	17	33	65	129	257	513	1025

4. **CLASS** Go over answers to problems 3–5. Ask students, "How is our research so far helping you think about how to design an effective plan to protect the village from the virus?"

2. RESEARCH THE PROBLEM: OUTBREAK!
RESEARCH PHASE 2: THERE'S A DOCTOR IN THE HOUSE

1. Imagine the same situation as in Research Phase 1, but now there is a doctor who can treat one person every day. What do you predict will happen as time goes on?

    _____

    _____

    _____

2. You will conduct an experiment as a class. Record the results of the experiment in Table 3.2 below.

Table 3.2: How Virus Spreads Through the Classroom When There Is a Doctor

TIME (DAYS)											
NUMBER OF SICK PEOPLE											

3. How did having the doctor in the classroom affect the spread of the virus over time? Look for patterns.

    _____

    _____

    _____

4. Represent the pattern(s) you found using pictures, symbols, or variables.

5. How many people do you think would be sick after 20 minutes? _____

    How did you determine this? _____

    _____

    _____

    _____

2. RESEARCH THE PROBLEM: OUTBREAK!
RESEARCH PHASE 3: EVERYONE HAS AN AIR-FILTRATION MASK

OBJECTIVES

Students will:
- participate in a simulation and collect simulation data
- identify and extend exponential patterns
- generalize and represent a pattern using symbols

MATERIALS

For the teacher:
- 1 lump of clay or modeling compound

For each student:
- 1 penny

1.   CLASS   Instruct students to read problem 1 together. After discussing students' predictions, follow these instructions to conduct the simulation:

- Note that there are no doctors in this simulation.

- Begin by giving each student 1 penny. This penny represents an air-filtration mask. With an air-filtration mask, a student has a 50% chance of getting sick. If a healthy person comes into contact with the virus, he or she will flip the coin. Heads = person gets sick; tails = person stays healthy.

- Start with time = 0 with 1 person sick. This person will split his or her clay and give half to a healthy person. The healthy person will flip his or her coin. Heads means the person keeps the clay and become "sick;" tails means the person returns the clay to the sick person and remains healthy. Keep track of the number of sick people after each time interval using a table on the board.

- At each time interval, each sick person will attempt to infect 1 healthy person. The healthy person will flip a coin to determine whether or not he or she gets sick. Continue simulation until all are sick.

- *Note:* During a time interval, sick people can't attempt to infect the same healthy person more than once. Once a healthy person flips tails once, he or she is safe for that entire turn.

- The time intervals are measured in "fast" days, which means that they only need to last as long as it takes to pass the virus once and record the number of sick people.

2.   **TEAMS** Instruct teams to answer questions 3 and 4.

ASK THE CLASS:

- If air masks work better than doctors, should we not have any doctors? Assuming that sick people cannot get better on their own without treatment from doctors, can a virus be contained over a long period of time?

  Possible Answer(s): Allow students to speculate and state their reasoning. It's fine at this point if they don't yet realize that while air masks, hand gel, and vaccinations will slow the spread by reducing the percentage of infection, if there are no doctors to turn sick people back into healthy people and assuming that no sick person recovers on his or her own, eventually everyone will become sick. If students don't come to this realization, they will later see this result when using the computer applet.

2. RESEARCH THE PROBLEM: OUTBREAK!
RESEARCH PHASE 3: EVERYONE HAS AN AIR-FILTRATION MASK

1. Imagine, once again, that the virus is spreading in the same way as in Research Phase 1. You no longer have a doctor. Now, each person in the classroom has an air-filtration mask. When a person is wearing the mask, he or she has a 50% chance of becoming sick when coming into contact with an infected person. What do you predict will happen as time goes on?

    _____

    _____

    _____

2. You will conduct an experiment as a class. Please record the results of the experiment in Table 3.3 below.

Table 3.3: How Virus Spreads Through the Classroom When Everyone Wears an Air Mask

TIME (DAYS)											
NUMBER OF SICK PEOPLE											

3. How did wearing the air-filtration masks affect the spread of the virus over time? Look for patterns.

    _____

    _____

    _____

4. Compare your results in Table 3.2 and Table 3.3. Which was more effective in slowing the spread of the virus, the doctor or the air-filtration masks?

    _____

    _____

    _____

Teacher Page

2. RESEARCH THE PROBLEM: OUTBREAK!
RESEARCH PHASE 4: GRAPHING THE DATA

OBJECTIVES

Students will:
- graph simulation data
- compare the results of the three simulations
- connect the results of research to the design challenge

MATERIALS

For each team:
- colored pencils (at least 3 different colors)

For each class:
- a sheet of chart paper for the Rules of Thumb list (see page 21)

1. **TEAMS** Instruct teams to graph their data from research phases 1–3 in different colors. Discuss which are the independent (*x*) and dependent (*y*) variables. If the class needs practice graphing or scaling the axes, use the activities on pages 5–13 as review or reinforcement. Provide the Rubric for Graphs on page 152 so students can assess their work.

2. **CLASS** Ask students to compare the graphs from the three simulations and describe the trends. Ask these questions to connect the research with the design challenge:

   - Do any of the simulations meet the design criteria? Explain.

   - How can the research so far help students select the most effective virus interventions for the lowest cost?

     Possible Answer(s): The research shows us that using the air-filtration masks is more cost-effective than hiring one doctor. Using masks slows the rate of the virus spread more than hiring one doctor and is cheaper than hiring one doctor. We can infer from this that interventions that reduce the infection rate for the whole village to 50% or less is more cost-effective than hiring one doctor.

Add students' responses to the Rules of Thumb list.

ASSESSMENT

Use the Rubric for Graphs on page 152 to assess students' graphs.

Name
STUDENT PAGE

RESEARCH

2. RESEARCH THE PROBLEM: OUTBREAK!

RESEARCH PHASE 4: GRAPHING THE DATA

Use the grid below to plot the results of your simulations from research phases 1 to 3. Use a different color to plot data from each phase. Label the axes and give the graph a title. Use the rubric provided by your teacher to assess your work.

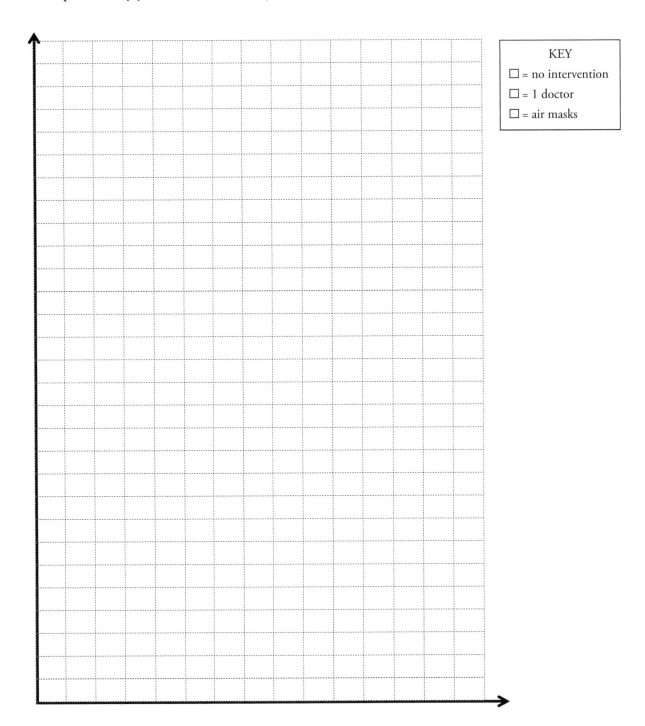

KEY

☐ = no intervention

☐ = 1 doctor

☐ = air masks

Teacher Page

2. RESEARCH THE PROBLEM: OUTBREAK!
RESEARCH PHASE 5: COMBINING INTERVENTIONS

OBJECTIVE: Students will calculate compound probabilities.

MATERIALS
For each individual or pair:
- 1 calculator

1. **CLASS** As you are reviewing the list of interventions on page 124 with the class, you may want to discuss this ethical issue: "Would it be right to vaccinate some but not all villagers? How could this be made fair?"

2. **CLASS** Do the first combination together to make sure students understand that they multiply the percentages (in decimal form) to get the percent chance for combined interventions.

3. **INDIVIDUALS** or **PAIRS** Assign individual students and/or pairs to complete calculating the percentages of infection for various combinations of interventions.

4. **CLASS** Go over student answers.

5. **CLASS** Run the simulation with and without vaccination and with and without air-filtration masks. Use a two-way table to look at bivariate data.

OPTIONAL CCSS ENHANCEMENT
To address additional aspects of the Common Core State Standards, reinforce the relationship between probabilities expressed as percents and as fractions when comparing the chances of a healthy person getting sick with various interventions.

---

2. RESEARCH THE PROBLEM: OUTBREAK!

RESEARCH PHASE 5: COMBINING INTERVENTIONS

INTERVENTION	EFFECT OF INTERVENTION	COST OF INTERVENTION
doctor	can treat 1 person per day	$1,600 per doctor
air-filtration masks	reduces chance of getting virus by 50%	$750 for a village supply
antiviral hand gel	reduces chance of getting virus to 25%	$1,000 for a village supply
vaccinations	reduces starting chance of getting virus to 4%	$120 for 1 villager

The table above shows the effects of each intervention used alone. What would be the effects if you combined them? Complete the table of combined interventions below. Round all answers to the nearest tenth of a percent. Then, write the percentage as a decimal number between 0 and 1.

INTERVENTIONS	CHANCE OF A HEALTHY PERSON GETTING SICK WITH THESE INTERVENTIONS	
	%	Decimal
1. air-filtration mask and antiviral hand gel		
2. air-filtration mask and vaccination		
3. antiviral hand gel and vaccination		
4. air-filtration mask, antiviral hand gel, and vaccination		

2. RESEARCH THE PROBLEM: OUTBREAK!
RESEARCH PHASE 6: DOCTORS ONLY!

This is an optional research phase that requires each team or pair of students to have access to a computer. It might be better for students to work in pairs rather than in teams of three or four in order to allow all students to actively participate in this research phase. The goal of this phase is to encourage students to systematically find a way to test plans to find a plan that is both successful and as low in cost as possible. The scenario is simple—only use doctors in the plan—which means that the infection rate is at 100%. Students will find that it takes a great number of doctors to keep the number of sick villagers to 10 or fewer. In fact, a successful plan might require 35 or more doctors.

OBJECTIVE: Students will use a computer model to meet the design criteria using only doctors.

MATERIALS
For the class:
- LCD projector and computer with doctors scenario applet

For each team or pair (preferred):
- access to a computer with doctors scenario applet
- 1 printed copy of doctors scenario applet readme file (optional)

BEFORE YOU TEACH
Take some time to become familiar with the doctors scenario applet. Read the readme file in the applet folder.

1. **CLASS** Explain to students that so far, they have only used physical models to represent the spread of a virus—that is, they used clay and coins. Today, they will use a computer model to represent the spread of a virus. The computer program will follow the same rules that students used during the class simulations. Read the first paragraph on page 128 and make sure that students understand the scenario and the challenge.

2. **TEAMS** Instruct students to make a prediction about how many doctors they would need to make a working plan and to explain their prediction in question 1.

ASK THE CLASS:
- Why do you think that we need to test each plan several times to determine whether or not it is successful?

   Possible Answer(s): We need to test each plan several times to ensure that we don't base the success of a plan on the results of just one trial.

- If a plan fails the first two times, do you need to continue testing the plan? Why or why not?

   Possible Answer(s): No, once a plan fails two times, it is not going to meet the criterion of four out of five successful trials.

Teacher Page *(continued)*

3.   **CLASS**

- To open the applet, open the Amazon Simulation 1 folder and click on **index.html.** The applet should load in a Java-enabled web browser. Click on **More Information** to bring up the readme file or open the readme file directly from the Simulation 1 folder. The readme file contains detailed instructions that you can print out and copy for students.

- Demonstrate the applet to the class by doing an example such as testing a plan with two doctors. Show them how to edit settings as you demonstrate the applet.

- To populate the board, students click on the reset button. They will notice that only the sick person and the healthy villagers appear on the board. This is because the doctors will appear after a short delay, similar to what happened in Research Phase 2.

- Run the simulation by clicking on **start.** Once the simulation finishes, click on **view graph** to see results.

- As the program runs, note changes in the total sick variable, and the number of sick increases and decreases.

- The simulation will automatically stop at the end of 30 "days." Then you can look at the graph.

- To restart the simulation with the same number of doctors, click on the reset button.

4.   **TEAMS**   Instruct students to follow the instructions in number 2 to find a successful plan. After finding a successful plan, they can then answer questions 3–5. Give students about 15 to 20 minutes to do this activity.

5. **CLASS** Debrief the activity by asking pairs or teams to share successful plans. Note any differences in results.

   ASK THE CLASS:

   - Who thinks that they have the lowest-cost plan that meets the criteria for success?

   - How did you find your lowest-cost and successful plan?

   - Why do some plans work for some teams but not others?

     Possible Answer(s): One successful run could be due to luck—the random movement of the sick person and the doctors may make it so that the sick person doesn't infect anyone before he or she is treated by one of the doctors. The reverse can be true—"unlucky" movements and early infection can rapidly spread the disease. But by testing the plan several times, one can see whether overall (or in this case four out of five times) the plan will be successful. A plan that succeeds five out of five times may still fail on the sixth trial or the 100th trial, so it might be a good opportunity to discuss the concept of "risk" with students.

   - Can one design a plan that works 100% of the time?

   - What if a plan works 99% of the time? Is it worth the risk of 1% failure to implement such a plan?

   - Four out of five successful trials is 80%. Is this percentage of success acceptable when people's health is at risk?

   - What would you consider an appropriate percentage of success?

6. **CLASS** Debrief questions 3–5. Encourage students to note how expensive it is to only use doctors to contain the virus and to consider the importance of reducing the infection rate using other interventions. In this simulation, as in real life, preventive public health measures are often more crucial to the success of containing infectious diseases than simply having lots of doctors to treat sick patients.

2. RESEARCH THE PROBLEM: OUTBREAK!
RESEARCH PHASE 6: DOCTORS ONLY!

In this research phase, you will use a computer model to simulate the spread of the virus and find the lowest-cost plan that still successfully meets the virus-containment criterion (the number of sick villagers must not go above 25% of the village in 30 days). In this scenario, you may only use doctors from the list of interventions. This means that the infection rate is 100%!

1. First, make a prediction: How many doctors do you think you will need to successfully contain the virus in this scenario? Explain your reasoning for your estimate.

   _____

   _____

2. Start by testing your prediction. Then adjust your estimate up or down depending on the test results. Test each plan several times (up to five times). Put a tally mark in the appropriate column as you test the same plan. Then circle YES if the plan works at least four out of five times.

NUMBER OF DOCTORS	COST OF PLAN	TRIALS THAT WORK	TRIALS THAT DIDN'T WORK	SUCCESSFUL PLAN?
				YES NO
				YES NO
				YES NO
				YES NO

3. What was the lowest-cost and most successful plan you could find? _____

4. What strategies did you use to find the lowest-cost and most successful plan?

   _____

   _____

5. Note how expensive it is to implement a successful plan that only uses doctors. What can you learn from your observations of how the flu spread in this scenario to help you design a successful plan that costs less than $10,000?

   _____

   _____

   _____

Teacher Page

3. BRAINSTORM POSSIBLE SOLUTIONS: OUTBREAK!

OBJECTIVES
Students will:
- review the design challenge's criteria and constraints
- individually choose interventions to include in their virus-containment plans and share plans with their teams

MATERIALS
For the class:
- the Rules of Thumb list

1. **CLASS** Review the design criteria and constraints on the next page.

2. **INDIVIDUALS** Instruct students to individually brainstorm some ideas for the virus-containment plan, write down their proposed plan, and complete question 1. Remind students that they can spend up to $10,000 but should challenge themselves to keep the cost as low as possible and still create an effective plan.

3. **TEAMS** Instruct students to take turns sharing their plans with their teammates. Each person should share his or her plan's cost, interventions chosen, rationale (why it's a good plan), and drawbacks (e.g., expense). While one person is sharing, the rest of the team should take notes and complete the table in question 2. After everyone has shared, team members may comment on one another's plans, saying what they like or don't like about each plan.

INTERESTING INFO
Vaccinations are small, controlled doses of a bacteria or virus that are injected in a patient to induce the immune system to build up resistance. This way, if later on a person is exposed to the same bacteria or virus, the immune system will know how to respond.

3. BRAINSTORM POSSIBLE SOLUTIONS: OUTBREAK!

You are now ready to design a plan to contain a virus spreading through the village of 40 people. Remember to keep these design requirements in mind!

ENGINEERING CRITERIA	
VIRUS CONTAINED →	The percent of villagers infected does not go above 25% for 30 days.
LOW COST →	The plan is as low in cost as possible. You have a maximum budget of $10,000, but try to find a lower cost plan that works.
HIGH CHANCE OF SUCCESS →	The plan must meet the virus-containment criterion at least four out of five trials.

ENGINEERING CONSTRAINTS		
Remember the interventions that are available for your plan:		
INTERVENTION	**EFFECT OF INTERVENTION**	**COST OF INTERVENTION**
doctor	can treat 1 person per day	$1,600 per doctor
vaccinations	reduces starting chance of getting virus to 4%	$120 for 1 villager
antiviral hand gel	reduces chance of getting virus to 25%	$1,000 for a village supply
air-filtration masks	reduces chance of getting virus by 50%	$750 for a village supply

INDIVIDUAL DESIGN

1. On your own, come up with one possible solution. Record the interventions you chose.

    - number of doctors: _____

    - number of vaccinations: _____

    - antiviral hand gel: ☐ yes   ☐ no

    - air-filtration masks: ☐ yes   ☐ no

    a. What is the cost of this solution? _____

    b. What makes this a good solution? What are its advantages?

    _____

    _____

    c. What are the drawbacks, or disadvantages, of this solution?

    _____

    _____

2. Each member of your team should now describe his or her idea to the group. As you discuss your ideas, fill in the chart below.

TEAM MEMBER NAME	COST OF SOLUTION	BENEFITS OF SOLUTION	DRAWBACKS OF SOLUTION

4. CHOOSE THE BEST SOLUTION: OUTBREAK!

OBJECTIVE: Students will decide on a team virus-containment plan.

TEAMS Instruct teams to discuss how to decide on one team plan (e.g., look for commonalities among all possible plans, choose the best parts of each team member's proposed plans, and combine them to create a team plan). Once they have decided on the team plan, students should answer questions 1–4.

Name

CHOOSE

4. CHOOSE THE BEST SOLUTION: OUTBREAK!

TEAM DESIGN

1. Consider the chart you filled out on page 131. As a group, choose the "best" solution. Record the interventions you chose.

    - number of doctors: _____

    - number of vaccinations: _____

    - antiviral hand gel: ☐ yes ☐ no

    - air-filtration masks: ☐ yes ☐ no

2. For your solution, what is the chance that healthy villagers will get sick if they come into contact with an infected person?

    _____

3. If you are giving out vaccinations to some but not all of the villagers, how will you decide which villagers will get the vaccinations? Explain your reasoning.

    _____

    _____

    _____

    _____

4. If you are hiring doctors, where should they be located within the shabono? Explain your reasoning.

    _____

    _____

    _____

    _____

Teacher Page

5. BUILD A PROTOTYPE/MODEL: OUTBREAK!

OBJECTIVE: Students will learn how to use the computer model to test their plans.

MATERIALS

For the class:
- LCD projector and computer with Amazon Simulation 2

For each team:
- printed copy of the readme file from the Amazon Applet 2 folder (optional)

1. **CLASS** Tell students that they will build a model of their designs using a computer virus simulation applet. Read the paragraph on page 135 that explains why one would use a computer model to test the virus-containment plan.

2. **CLASS** Demonstrate the applet to the class using an LCD projector.
   - Open the Amazon Simulation 2 folder and click on **index.html** to open the applet in a web browser.
   - Show students how to open the readme file to get directions on using the applet by clicking on **More Information,** or distribute a printed copy of the readme file to use as reference.

BUILD

5. BUILD A PROTOTYPE/MODEL: OUTBREAK!

You do not "build" anything in this design step because you will use a computer simulation model to test your plan. In this model, each villager and any doctors move randomly on a roughly circular background that is 16 units in diameter. When a healthy villager and a sick villager are in adjacent squares, the healthy villager has a percent chance of getting sick according to your intervention plan. Doctors can successfully treat one sick villager in one of the adjacent squares per day. Computer models use the speed of computers to easily and quickly execute complex rules. They also allow users to manipulate variables in the model to test what happens in different scenarios.

Teacher Page

6. TEST YOUR SOLUTION: OUTBREAK!

OBJECTIVE: Students will test their teams' virus-containment plans.

MATERIALS

For the class:
- LCD projector and computer with Amazon Simulation 2 folder

For each team or pair (preferred):
- access to a computer with Amazon Simulation 2 folder
- 1 printed copy of Amazon Applet 2 readme file (optional)

BEFORE YOU TEACH

Take some time to become familiar with the doctors scenario applet. Read the readme file in the applet folder.

1. **PAIRS** Instruct students to work in pairs to complete the Notes table on page 138 by using their answers from Research Phase 5.

2. **CLASS** Demonstrate how to complete the Original Plan table by using a sample like the one below:

   - number of doctors: ___2___

   - number of vaccinations: ___20___

   - antiviral hand gel: ✓ Yes ☐ No

   - air-filtration masks: ☐ Yes ✓ No

   - Show students how to use the infection rates table to find the appropriate column according to the plan.

3. **TEAMS** Instruct teams to complete the Original Plan table according to their chosen team plans. Walk around and check the tables to make sure that they are correctly done. Then teams can test their plans. If they finish early, they can move on to Challenge 1 and Challenge 2 on page 139.

   Challenge 1 is for teams whose plans did not work, and Challenge 2 is for teams to optimize their working plan to make it more cost-effective. Provide the Rubric for Test, Communicate, and Redesign Steps on page 158 so students can assess their work.

> **TIP**
> As an incentive, offer a small prize (e.g., candy or school supplies) to the team that can find the most successful low-cost plan.

© Museum of Science (Boston), Wong, Brizuela

OBJECTIVE: Students will optimize their teams' virus-containment plans to be as cost-effective as possible.

1. **TEAMS** After adjusting their original plan, students fill out the corresponding variables table on page 139 before entering their values on the applet.

2. **CLASS**
   ASK THE CLASS:
   - Which teams found a successful plan?
   - What was the cost of your plan?
   - Which team found the most successful low-cost plan? What were the interventions of this plan? How does the plan compare with other successful but more expensive plans?
   - Would you choose a more expensive plan that works five out of five times or a cheaper plan that works four out of five times?
   - Is an 80% success rate sufficient? What would you consider an acceptable success rate?

It is possible to find a successful plan without using any doctors (that is, slow the infection rate enough so that fewer than 10 people are infected in the first 30 days). In the long run, however, if there is no doctor (and assuming no one gets better "naturally"), the entire village will get sick, no matter how low the infection rate.

6. TEST YOUR SOLUTION: OUTBREAK!

You may use the rubric provided by your teacher to assess your work on the next few pages.

ORIGINAL PLAN

number of doctors: _____   antiviral hand gel: ☐ yes ☐ no
number of vaccinations: _____   air-filtration masks: ☐ yes ☐ no

	TRIAL NUMBER					
	1	2	3	4	5	Did the plan work 4 out of 5 times?
Number of sick people at the end of 30 days						☐ yes ☐ no

NOTES

1. Complete the table below using your answers from Research Phase 5. Find the infection rate for vaccinated and unvaccinated villagers by clicking "Edit Settings."

	NO INTERVENTION	MASK ONLY	HAND GEL ONLY	MASK AND HAND GEL
Unvaccinated infection rate				
Vaccinated infection rate				

Challenge 1: If your original plan did not work, discuss how you can redesign your plan to make it work. Test your plan and record your results. Use the following table to determine the variables for your new plan and to test your new plan.

Challenge 2: If your original plan works, redesign your plan to find a lower-cost plan.

CHANGES TO YOUR ORIGINAL PLAN
Redesigned Plan 1

number of doctors: _____   antiviral hand gel: ☐ yes ☐ no
number of vaccinations: _____   air-filtration masks: ☐ yes ☐ no

	TRIAL NUMBER					
	1	2	3	4	5	Did the plan work 4 out of 5 times?
Number of sick people at the end of 30 days						☐ yes ☐ no

Redesigned Plan 2

number of doctors: _____   antiviral hand gel: ☐ yes ☐ no
number of vaccinations: _____   air-filtration masks: ☐ yes ☐ no

	TRIAL NUMBER					
	1	2	3	4	5	Did the plan work 4 out of 5 times?
Number of sick people at the end of 30 days						☐ yes ☐ no

Redesigned Plan 3

number of doctors: _____   antiviral hand gel: ☐ yes ☐ no
number of vaccinations: _____   air-filtration masks: ☐ yes ☐ no

	TRIAL NUMBER					
	1	2	3	4	5	Did the plan work 4 out of 5 times?
Number of sick people at the end of 30 days						☐ yes ☐ no

7. COMMUNICATE YOUR SOLUTION: OUTBREAK!

OBJECTIVE: Students will answer questions to reflect on their designs and discuss the results of their tests.

1.　TEAMS　Teams should discuss and answer questions 1–5 on the next page.

2.　CLASS　Discuss students' answers.

COMMUNICATE

7. COMMUNICATE YOUR SOLUTION: OUTBREAK!

1. Do you think that your plan to contain a virus was successful? Did it meet all of the criteria and constraints? Explain.

    _____

    _____

    _____

2. Specifically, what are some of the strengths or advantages of your plan? Explain.

    _____

    _____

    _____

3. What are some drawbacks or disadvantages of your plan? Explain.

    _____

    _____

    _____

4. If you had an unlimited amount of money, what would you do differently? How would you change your plan?

    _____

    _____

    _____

5. How would your plan work if the village was larger? Do you think it would be successful for a village of 100 people? 200 people? 300 people? Explain your reasoning.

    _____

    _____

    _____

Be prepared to present your answers to questions 1–5 to the class.

8. REDESIGN AS NEEDED: OUTBREAK!

OBJECTIVE: Students will answer questions to consider how they can redesign their virus-containment plans.

1. **TEAMS** Instruct teams to use what they learned from other teams' plans—both successes and failures—to improve their plans by answering questions 1 and 2 on the next page.

2. **CLASS** Wrap up the design challenge with the following questions:
   - How did you use math to solve the virus-containment design challenge?

     Possible Answer(s): We looked for patterns in our experiments to simulate the spread of a virus under different conditions and generalized the patterns. We also graphed the experiment results to help us look for patterns, used symbols to represent patterns, and calculated the percentage of infection given different combinations of interventions.

   - What are some techniques for finding and generalizing patterns?

     Possible Answer(s): Look for patterns in how the numbers are increasing or decreasing. Find differences between successive items in the sequence. Compare the sequence to known sequences. Look for a relationship between the position of each item (the time interval associated with the corresponding number of sick people) and the value of the item (number of sick people). Graph the data and use the shape of the line to determine whether the data is increasing linearly, exponentially, and so forth.

REDESIGN

8. REDESIGN AS NEEDED: OUTBREAK!

1. Based on the test of your virus-containment plan, what changes could you make to improve your plan?

_____

_____

_____

_____

Explain how these changes would improve your virus-containment plan.

_____

_____

_____

_____

2. Identify one thing that you learned from another group's virus-containment plan that you can use to improve your virus-containment plan.

_____

_____

_____

_____

Explain how this will improve your virus-containment plan.

_____

_____

_____

_____

INDIVIDUAL SELF-ASSESSMENT RUBRIC: OUTBREAK!

OBJECTIVE: Students will use a rubric to individually assess their involvement and work in this design challenge.

> **ASSESSMENT**
>
> Assign this reflection exercise as homework. You can write your comments on the lines below the self-assessment and/or use this in conjunction with the Student Participation Rubric on page 159.

TEAM EVALUATION: OUTBREAK!

OBJECTIVE: Students will evaluate and discuss how well they worked in teams.

> **ASSESSMENT**
>
> 1. **INDIVIDUALS** Assign this reflection exercise as homework or during quiet classroom time.
> 2. **TEAMS** Instruct students to share their team evaluation reflections with one another.
> 3. **CLASS** Point out any good examples of teamwork.

INDIVIDUAL SELF-ASSESSMENT RUBRIC: OUTBREAK!

Use this rubric to reflect on how well you met behavior and work expectations during this activity. Check the box next to each expectation that you successfully met.

LEVEL 1	LEVEL 2	LEVEL 3	LEVEL 4	BONUS POINTS
Beginning to meet expectations	Meets some expectations	Meets expectations	Exceeds expectations	
☐ I was willing to work in a group setting.	☐ I met all of the Level 1 requirements.	☐ I met all of the Level 2 requirements.	☐ I met all of the Level 3 requirements.	☐ I helped resolve conflicts on my team.
☐ I was respectful and friendly to my teammates.	☐ I recorded the most essential comments from other group members.	☐ I made sure that my team was on track and doing the tasks for each activity.	☐ I helped my teammates understand the things that they did not understand.	☐ I responded well to criticism.
☐ I listened to my teammates and let them fully voice their opinions.	☐ I read all instructions.	☐ I listened to what my teammates had to say and asked for their opinions throughout the activity.	☐ I was always focused and on task: I didn't need to be reminded to do things; I knew what to do next.	☐ I encouraged everyone on my team to participate.
☐ I made sure we had the materials we needed and knew the tasks that needed to be done.	☐ I wrote down everything that was required for the activity.	☐ I actively gave feedback (by speaking and/or writing) to my team and other teams.	☐ I was able to explain to the class what we learned and did in the activity.	☐ I encouraged my team to persevere when my teammates faced difficulties and wanted to give up.
	☐ I listened to instructions in class and was able to stay on track.	☐ I completed all the assigned homework.		☐ I took advice and recommendations from the teacher about improving team performance and used feedback in team activities.
	☐ I asked questions when I didn't understand something.	☐ I was able to work on my own when the teacher couldn't help me right away.		☐ I worked with my team outside of the classroom to ensure that we could work well in the classroom.
		☐ I completed all the specified tasks for the activity.		

Approximate your level based on the number of checked boxes: _____ Bonus points: _____

Teacher comments:

TEAM EVALUATION: OUTBREAK!

How well did your team work together to complete the design challenge? Reflect on your teamwork experience by completing this evaluation and sharing your thoughts with your team. Celebrate your successes.

RATE YOUR TEAMWORK. On a scale of 0–3, how well did your team do? 3 is excellent, 0 is very poor. Explain how you came up with that rating. Was it the same, better, or worse than the last activity?

LIST THINGS THAT WORKED WELL. Example: We got to our tasks right away and stayed on track.

LIST THINGS THAT DID NOT WORK WELL. Example: We argued a lot and did not come to a decision that everyone could agree on.

HOW CAN YOU IMPROVE TEAMWORK? Make the action steps concrete. Example: We need to learn how to make decisions better. Therefore, I will listen and respond without raising my voice.

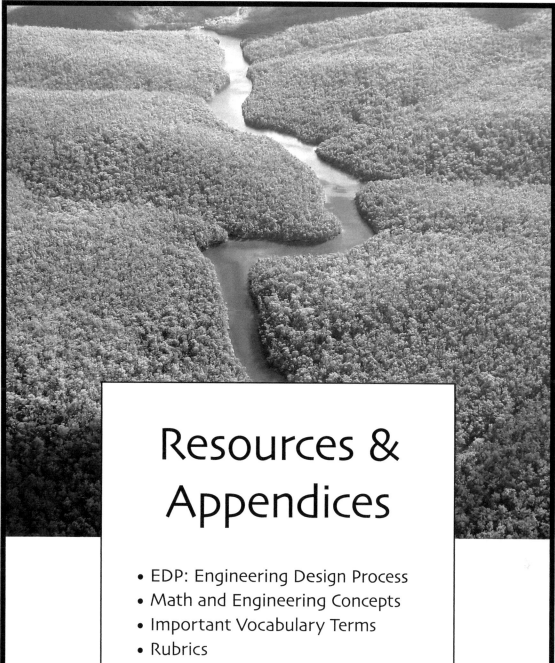

Resources & Appendices

- EDP: Engineering Design Process
- Math and Engineering Concepts
- Important Vocabulary Terms
- Rubrics
- Student Work Samples
- Appendix: Net of Rectangular Prism

EDP: ENGINEERING DESIGN PROCESS

Engineers all over the world have one thing in common. They use the engineering design process (EDP) to solve problems. These problems can be as complicated as building a state-of-the-art computer or as simple as making a warm jacket. In both cases, engineers use the EDP to help solve the problems. Although engineers may not strictly follow every step of the EDP in the same order all the time, the EDP serves as a tool that helps to guide engineers in their thinking process and approach to a problem. Below is a brief outline of each step.

DEFINE	The first step is to **define** the problem. In doing so, remember to ask questions! What is the problem? What do I want to do? What specifications should my solution meet to successfully solve the problem (also called "criteria")? What factors may limit possible solutions to this problem (also called "constraints")?
RESEARCH	The next step is to conduct **research** on what can be done to solve the problem. What are the possible solutions? What have others already done? Use the Internet and the library to conduct investigations and talk to experts to explore possible solutions.
BRAINSTORM	**Brainstorm** ideas and be creative! Think about possible solutions in both two and three dimensions. Let your imagination run wild. Talk with your teacher and fellow classmates.

4 CHOOSE	**Choose** the best solution that meets all the criteria and constraints. Any diagrams or sketches will be helpful for later engineering design steps. Make a list of all the materials the project will need.
5 BUILD	Use your diagrams and list of materials as a guide to **build** a model or prototype of your solution.
6 TEST	**Test** and evaluate your prototype. How well does it work? Does it satisfy the engineering criteria and constraints?
7 COMMUNICATE	**Communicate** with your peers about your prototype. Why did you choose this design? Does it work as intended? If not, what could be fixed? What were the trade-offs in your design?
8 REDESIGN	Based on information gathered in the testing and communication steps, **redesign** your prototype. Keep in mind what you learned from one another in the communication step. Improvements can always be made!

MATH AND ENGINEERING CONCEPTS

In these three *Amazon Mission* activities, you will integrate engineering with math to solve problems and design prototypes. Specifically, you will:

- collect, represent, and analyze nonlinear data
- represent exponential functions graphically and algebraically
- solve problems involving proportions and scaling, and build scale models
- solve problems involving rates and unit conversions
- use formulas to calculate geometric measurements of a sphere
- draw nets, and calculate the surface area of three-dimensional objects
- apply the engineering design process to solve problems

IMPORTANT VOCABULARY TERMS

AMAZONIA
the Amazon River Basin region, located in northern South America

ANTIMALARIAL MEDICINE
a medicinal drug used to treat and cure a person suffering from malaria by destroying the malaria parasite that is in the sick person's body

ENDANGERED
in danger of becoming extinct

ENGINEERING
the applications of math and science to practical ends, such as design or manufacturing

ENGINEERING CONSTRAINTS
limiting factors to consider when designing a model

ENGINEERING CRITERIA
specifications met by a successful solution

EXTINCT
no longer existing or living

INHABIT
to live, reside, or dwell in

INDIGENOUS
originating from or native to a region

MALARIA

a disease transmitted to humans by a bite from a female anopheles mosquito; may cause fever, chills, fatigue, confusion, anemia, and even coma or death

MERCATRON

a fictional, spherical object coated with a mercury-absorbing material

MERCURY

a silvery-white, poisonous metallic element

PROTOTYPE

a test model that contains only the essential features of the design, and serves as a basis or standard for later stages of the design

RAIN FOREST

a forest in a tropical region that has heavy annual rainfall

SCALE

a proportion used as a constant relationship between the dimensions of a model and the real object it represents

SCALE MODEL

an object that has been built to represent another, usually larger, object (the model is the same shape and has the same proportions, but is not the same size)

SHABONO

a large, wooden, circular structure surrounding an open area, which houses an entire Yanomami village

SPHERE

a three-dimensional geometric figure, shaped like a perfectly round ball

VIRUS

a microorganism that can only grow and multiply in the living cells of hosts such as humans, animals, and plants, and can cause diseases to spread in its hosts

YANOMAMI

an ethnic group native to the rain forests of Venezuela and Brazil

RUBRIC FOR GRAPHS: MALARIA MELTDOWN!, MERCURY RISING!, AND OUTBREAK!

	EXPERT (4)	COMPETENT (3)	BEGINNER (2)	NOVICE (1)
Labeling	☐ Clearly and appropriately labels the x- and y-axes ☐ Labels graph with a title that correctly identifies the data being represented ☐ Correctly identifies independent and dependent variables	☐ Appropriately labels the x- and y-axes ☐ Labels graph with a title that correctly identifies the data being represented ☐ Correctly identifies independent and dependent variables	☐ Leaves out one of the axis labels or inappropriately labels one axis (e.g., forgets to include units) ☐ Leaves out axis title or uses a title that partially identifies the data being represented ☐ Does not correctly identify independent and dependent variables	☐ Leaves out both axis labels or inappropriately labels both axes ☐ Leaves out title or uses title that does not identify the data being represented ☐ Does not correctly identify independent and dependent variables
Scales and intervals	☐ Uses scales and intervals for x- and y-axes that show the entire range of data; data is well spread out ☐ Shows equal intervals	☐ Uses scales and intervals for x- and y-axes that show the entire range of data, but data may not be well spread out ☐ Shows mostly equal intervals (one or two minor errors)	☐ Scales on x- and y-axes may not reflect the range of data needed, or intervals may not be appropriate to precisely show data ☐ Shows some equal intervals (a few major errors on one or both axes)	☐ Missing scale on one or more axes, scales on x- and y-axes do not reflect the range of data needed, or intervals are not appropriate to show the data ☐ Shows unequal intervals (e.g., uses data values as intervals)
Data Representation	☐ Graph accurately reflects data in the data table	☐ Graph mostly reflects data in the data table (one or two minor errors)	☐ Graph reflects some of the data in the data table (many minor errors or a few major errors)	☐ Graph reflects few to none of the data in the data table (major errors)
Type of graph	☐ Type of graph is appropriate for the kind of data	☐ Type of graph is appropriate for the kind of data	☐ Type of graph may not be appropriate for the kind of data	☐ Type of graph is not suitable to represent the kind of data

RUBRIC FOR EXPERIMENT DESIGN (STEP 2): MERCURY RISING!

	EXPERT (4)	COMPETENT (3)	BEGINNER (2)	NOVICE (1)
Overall completeness	☐ Selects a factor to test and gives rationale for why that factor can significantly affect the flow rate ☐ Discusses and writes out a set of test procedures ☐ Uses the procedures to test the factor ☐ Creates and completes a data table ☐ Creates a graph of the data ☐ Analyzes graph and comes up with a rule of thumb based on test results	☐ Completes at least 5 of the 6 parts	☐ Completes at least 3 of the 6 parts	☐ Completes fewer than 3 of the 6 parts
Quality of test procedures	☐ Designs test procedures that clearly and explicitly test a single variable while keeping all other conditions the same ☐ Writes out detailed step-by-step procedures that can be easily understood and followed ☐ Selects at least four to five variables that are appropriately spaced in value ☐ Creates clearly labeled data collection table ☐ Uses data labels that are appropriate to the factor being tested ☐ Carefully follows the testing procedures, and collects and records precise data in the data table	☐ Designs test procedures that test for a single variable, but may not explicitly explain how the variable is controlled ☐ Writes out step-by-step testing procedures ☐ Selects at least three to four variables that are mostly appropriately spaced in value ☐ Creates data collection table with data labels that are relevant to the factor being tested ☐ Follows most of the testing procedures, and collects and records data in the data table	☐ Designs test procedures, but may not design a controlled experiment that tests only one variable ☐ Writes out testing procedures that may be incomplete or missing key procedures ☐ Selects at least two to three variables, but some may not be appropriately spaced in value ☐ Creates a partial data collection table ☐ Uses some inappropriate data labels or leaves out some data labels ☐ Follows some of the testing procedures ☐ Collects and records some data in the data table	☐ Designs test procedures that are not controlled and does not test only one variable ☐ Does not write out testing procedures, or procedures are very vague and sketchy ☐ Does not create a data collection table ☐ Does not select variables that are appropriately spaced in value, or only tests one or two variables ☐ Does not use appropriate data labels or any labels ☐ Does not follow testing procedures ☐ Incorrectly records data or does not collect data

RUBRIC FOR EXPERIMENT DESIGN (STEP 2): MERCURY RISING! (CONTINUED)

	EXPERT (4)	COMPETENT (3)	BEGINNER (2)	NOVICE (1)
Graph completeness and quality	☐ Clearly and appropriately labels the x- and y-axes ☐ Labels graph with appropriate title ☐ Uses appropriate scales for x- and y-axes ☐ Graph accurately reflects data in the data table ☐ Type of graph is appropriate to the kind of data	☐ Appropriately labels the x- and y-axes ☐ Labels graph with appropriate title ☐ Uses mostly appropriate scales for x- and y-axes (may not use the most appropriate intervals) ☐ Graph mostly reflects data in the data table (one or two minor errors) ☐ Type of graph is appropriate to the kind of data	☐ Leaves out one of the axis labels or inappropriately labels one axis (e.g., forgets to include units) ☐ Leaves out title or uses inappropriate title ☐ Scales on x- and y-axes may not reflect the range of data needed, or intervals may not be appropriate to precisely show data ☐ Graph reflects some of the data in the data table (many minor errors or a few major errors) ☐ Type of graph may not be appropriate to the kind of data	☐ Leaves out both axis labels or inappropriately labels both axes ☐ Leaves out title or uses inappropriate title ☐ Missing scale on one or more axes, scales on x- and y-axes do not reflect the range of data needed, or intervals are not appropriate to show the data
Data analysis	☐ Thoughtfully analyzes data to draw conclusions that are directly relevant to successfully meeting design criteria and constraints ☐ Supports conclusion with evidence from the experiment results OR thoughtfully explains how data results are insufficient to support drawing any conclusions ☐ Shows a reasonable amount of skepticism in own results	☐ Analyzes data to draw conclusions that help students meet design criteria and constraints ☐ Supports conclusion with evidence from the experiment results OR explains how data results may be insufficient evidence to support drawing any conclusions	☐ Draws conclusions that somewhat help students meet design criteria and constraints ☐ Does not support conclusions with evidence from the experiment OR states that data results are not able to support drawing any conclusions but explains the reason in vague terms	☐ Draws conclusions that are not relevant to helping students meet design criteria and constraints ☐ Does not support conclusions with evidence from the experiment OR states that data results cannot support drawing any conclusions but does not explain why

© Museum of Science (Boston), Wong, Brizuela

RUBRIC FOR ENGINEERING DRAWINGS (STEP 4): MALARIA MELTDOWN!

	EXPERT (4)	COMPETENT (3)	BEGINNER (2)	NOVICE (1)
Quality of idea	☐ Selects a design that addresses the problem ☐ Uses materials efficiently and with purpose ☐ Chooses design as a team through thoughtful deliberation ☐ Is able to express rationale for each part of the design	☐ Selects a design that addresses most of the problem ☐ Uses material mostly efficiently and with purpose ☐ Chooses design as a team after some deliberation ☐ Is able to express rationale for most parts of the design	☐ Selects a design that somewhat addresses the problem ☐ Uses materials with some purpose ☐ Chooses design as a team after a little discussion ☐ Is able to express rationale for some parts of the design	☐ Selects a design that does not address the problem ☐ Does not use materials with purpose ☐ Chooses design hastily without much discussion ☐ Is not able to express rationale for design
Communication	☐ Accurately draws a 3-D representation of the carrier ☐ Accurately draws a net of the carrier ☐ Labels all dimensions ☐ Uses appropriate units ☐ Labels all the materials used in the design	☐ Draws a 3-D representation of the carrier with minor errors ☐ Draws a net of the carrier with minor errors ☐ Labels most dimensions ☐ Uses appropriate units ☐ Labels most of the materials used in the design	☐ Attempts to draw a 3-D representation of the carrier ☐ Attempts to draw a net of the carrier ☐ Labels some dimensions ☐ Uses somewhat appropriate units ☐ Labels some of the materials used in the design	☐ Missing one or both 3-D representation and net drawing of the carrier ☐ Does not label dimensions. ☐ Units are not included or are inappropriate ☐ Does not label materials used in the design

Name

RUBRIC FOR ENGINEERING DRAWINGS (STEP 4): MERCURY RISING!

	EXPERT (4)	COMPETENT (3)	BEGINNER (2)	NOVICE (1)
Quality of idea	☐ Selects a design that addresses the problem ☐ Uses materials efficiently and with purpose ☐ Chooses design as a team through thoughtful deliberation ☐ Is able to express rationale for each part of the design	☐ Selects a design that addresses most of the problem ☐ Uses material mostly efficiently and with purpose ☐ Chooses design as a team after some deliberation ☐ Is able to express rationale for most parts of the design	☐ Selects a design that somewhat addresses the problem ☐ Uses materials with some purpose ☐ Chooses design as a team after a little discussion ☐ Is able to express rationale for some parts of the design	☐ Selects a design that does not address the problem ☐ Does not use materials with purpose ☐ Chooses design hastily without much discussion ☐ Is not able to express rationale for design
Communication	☐ Draws a detailed design of the filter, including the number of holes, size of holes, where the water goes in and out, and where the Mercatrons are located ☐ Labels the functions of all parts of the design. ☐ Labels all the materials used in the design	☐ Draws a mostly detailed design of the filter, missing one of the following: number of holes, size of holes, where the water goes in and out, or where the Mercatrons are located ☐ Labels the functions of most parts of the design ☐ Labels most of the materials used in the design	☐ Draws a partial design of the filter, missing two of the following: number of holes, size of holes, where the water goes in and out, and where the Mercatrons are located ☐ Labels the functions of some parts of the design ☐ Labels some of the materials used in the design	☐ Draws a rough design of the filter; missing three or more of the following: number of holes, size of holes, where the water goes in and out, and where the Mercatrons are located ☐ Does not label the functions of the parts of the design ☐ Does not label materials used in the design

RUBRIC FOR PROTOTYPE/MODEL (STEP 5): MALARIA MELTDOWN!, MERCURY RISING!, AND OUTBREAK!

	EXPERT (4)	COMPETENT (3)	BEGINNER (2)	NOVICE (1)
Completeness	☐ Builds a model that meets all criteria and constraints ☐ Follows the design sketch ☐ Follows cleanup procedures	☐ Builds a model that addresses most of the criteria and constraints ☐ Follows most of the design sketch ☐ Follows cleanup procedures	☐ Builds a model that addresses some of the criteria and constraints ☐ Follows some of the design sketch ☐ Partially follows cleanup procedures	☐ Builds an incomplete model ☐ Does not follow the design sketch ☐ Does not follow cleanup procedures
Craftsmanship	☐ Takes care in constructing model; is adept with tools and resources, and makes continual adjustments to optimize the model/prototype ☐ Demonstrates persistence with minor problems	☐ Uses tools and resources with little or no guidance ☐ Refines model to enhance appearance and capabilities	☐ Uses tools and resources with some guidance; may have difficulty selecting the appropriate resource ☐ Refines work, but may prefer to leave model as first produced	☐ Needs guidance in order to use resources safely and appropriately ☐ Model/prototype is crude, with little or no refinements made

RUBRIC FOR TEST, COMMUNICATE, AND REDESIGN STEPS: MALARIA MELTDOWN!, MERCURY RISING!, AND OUTBREAK!

	EXPERT (4)	COMPETENT (3)	BEGINNER (2)	NOVICE (1)
Completeness	☐ Carefully follows the testing procedures and documents all testing results	☐ Follows the testing procedures and documents most of the testing results	☐ Follows some of the testing procedures and documents some of the testing results	☐ Does not follow the testing procedures and does not document the testing results
Model performance	☐ The model fully meets the design constraints and criteria.	☐ The model meets most of the design constraints and criteria.	☐ The model meets some design constraints and criteria completely but ignores others.	☐ The model fails to meet design criteria and constraints.
Quality of reflection	☐ Specific improvement ideas are generated and documented	☐ Some general improvement ideas are generated and documented.	☐ The need for improvements is recognized and some ideas are generated, but documentation is not complete.	☐ Little interest is taken in improving the prototype or model, despite problems detected during testing. There is no evidence of inclination or ability to generate refinement solutions.

STUDENT PARTICIPATION RUBRIC

	0 POINTS JUST BEGINNING	1 POINT	2 POINTS	3 POINTS	SCORE
CONTENT CONTRIBUTION					
Sharing information	Discussed very little information related to the topic	Discussed some basic information; most related to the topic	Discussed a great deal of information; all related to the topic	Discussed a great deal of information showing in-depth analysis and thinking skills	
Creativity	Did not contribute any new ideas	Contributed some new ideas	Contributed many new ideas	Contributed a great deal of new ideas	
RESPONSIBILITY					
Completion of assigned duties	Did not perform any assigned duties	Performed very few assigned duties	Performed nearly all assigned duties at the level of expectation	Performed all assigned duties; did extra duties	
Attendance	Was never present or was always a negative influence when present	Attended some group meetings; absence(s) hurt the group's progress	Attended most group meetings; absence(s) did not affect group's progress or made up work	Attended all focus group meetings	
Staying on task	Not productive during group meetings; often distracted the team	Productive some of the time; needed reminders to stay on task	Productive most of the time; rarely needed reminders to stay on task	Used all of the focus group time effectively; productive at all times	
TEAMWORK					
Cooperating with teammates	Was rarely talking or always talking; usually argued with teammates	Usually did most of the talking, rarely allowing others to speak; sometimes argued	Listened, but sometimes talked too much; rarely argued	Listened and spoke a fair amount; never argued with teammates	
Making fair decisions	Always needed to have things his or her way; easily upset	Usually wanted to have things his or her way or often sided with friends instead of considering all views	Usually considered all views	Always helped team to reach a fair decision	
Leadership	Never took lead; needed to be assigned duties	Took a lead at least once; volunteered for duty	Took a lead more than once; volunteered for duties and helped others	Played essential role in organizing the group; frequently took lead; always helped others	

Teacher: _____   Total score: _____ / 24

STUDENT WORK SAMPLE 1

MALARIA MELTDOWN! STEP 4: ENGINEERING DRAWING

Use the rubric provided by your teacher to assess the following student work sample. Write a brief explanation for the grade you assign and how the work can be improved.

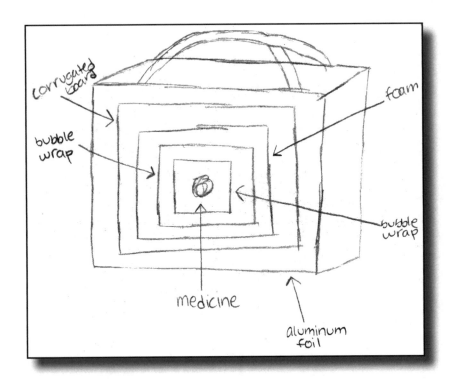

1. Grade: _____

2. Reasons for grade: _____

   _____

   _____

   _____

3. How work can be improved: _____

   _____

   _____

   _____

STUDENT WORK SAMPLE 2
MALARIA MELTDOWN! STEP 4: ENGINEERING DRAWING

Use the rubric provided by your teacher to assess the following student work sample. Write a brief explanation for the grade you assign and how the work can be improved.

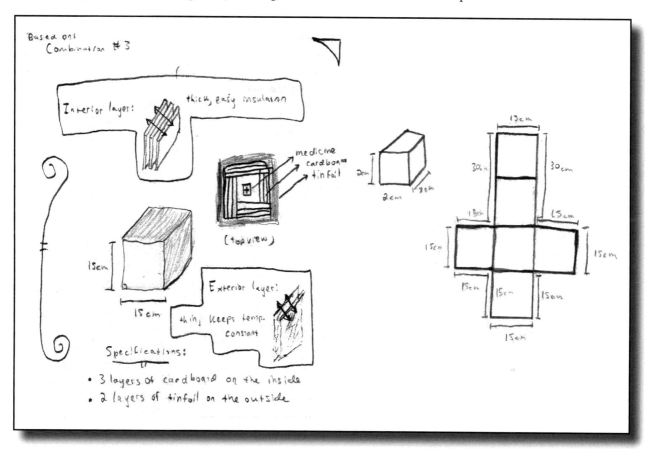

1. Grade: _____

2. Reasons for grade: _____

   _____

   _____

   _____

3. How work can be improved: _____

   _____

   _____

   _____

STUDENT WORK SAMPLE 3
MALARIA MELTDOWN! STEP 4: ENGINEERING DRAWING

Use the rubric provided by your teacher to assess the following student work sample. Write a brief explanation for the grade you assign and how the work can be improved.

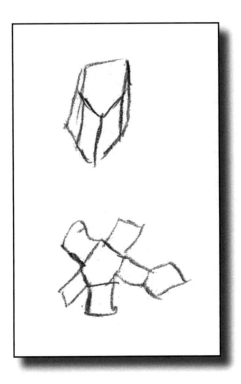

1. Grade: _____

2. Reasons for grade: _____
    _____
    _____
    _____

3. How work can be improved: _____
    _____
    _____
    _____

STUDENT WORK SAMPLE 4
MALARIA MELTDOWN! STEP 4: ENGINEERING DRAWING

Use the rubric provided by your teacher to assess the following student work sample. Write a brief explanation for the grade you assign and how the work can be improved.

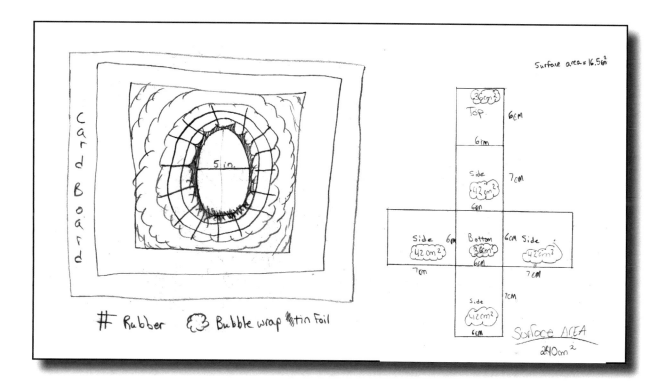

1. Grade: _____

2. Reasons for grade: _____

    _____

    _____

    _____

3. How work can be improved: _____

    _____

    _____

    _____

NET OF A RECTANGULAR PRISM

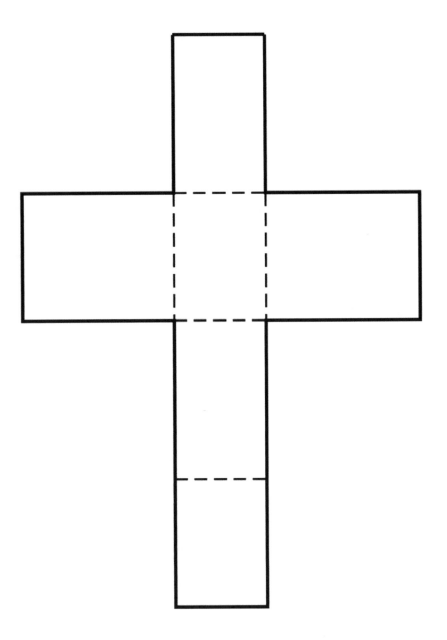

Answer Key

ANSWER KEY

LINE GRAPH ACTIVITY

Exercise 1

1. Mileage; it is an "input."
2. Value; it is an "output."
3. *x* mileage (independent variable)
4. *y* value (dependent variable)
5. a. 0–120,000; 120,000
   b. 25
   c. 120,000, 25; data range is greater than number of boxes so 120,000 ÷ 25 = 4,800
   d. Every box is worth 4,800.
   e. *x*-axis should be labeled "Mileage" with appropriate units.

6–7. See sample graph; answers will vary.

Sample graph:

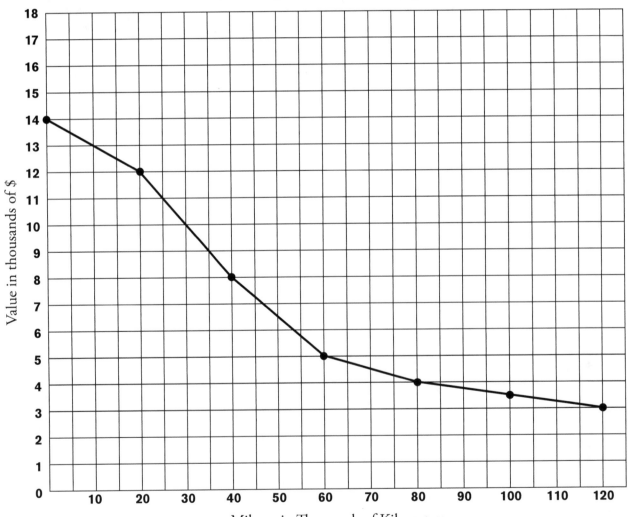

Relationship Between Truck Mileage and Value

8. Answers will vary.
9. $3,500
10. 20,000–40,000 kilometers
11. $6,500
12. $2,500, by extending the line or estimating the next decrease based on the previous decrease

Sample graph:

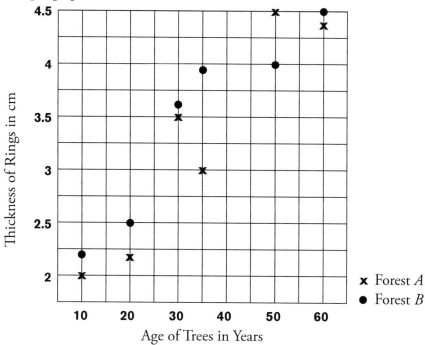

Age of Trees in Years

Exercise 2
1. 2.5 cm
2. 30 or 60 years old
3. 5.0 cm, based on the general trend (up). However, this is not guaranteed due to the unpredictability of climate and weather.
4. Rainfall, for example, has been more irregular in Forest *A* than in Forest *B*. The pattern of growth is more erratic in Forest *A* than Forest *B*.

CONVERTING UNITS ACTIVITY
1. 10 m per 3.6 seconds
2. 5 meters per 3.6 seconds
3. 50 m per 2.77 minutes
4. 100 mL per 346 seconds

REPRESENTING PATTERNS ACTIVITY
1. a. # squares = s
      $s_{next} = s_{current} + 4$
   b. # squares = (Figure # + 1) × 4
      50th figure = (50 + 1) × 4 = 51 × 4 = 204
   c. $4(n + 1) = s$
2. a. $x \bullet 3 = y$
   b. $(x \bullet 3) - 1 = y$
   c. $x^2 = y$
   d. $3^x = y$

INTRODUCING THE ENGINEERING DESIGN PROCESS (EDP)
1. Answers will vary.
2. Step 1: Define the problem.
   Step 2: Research the problem.
   Step 3: Brainstorm possible solutions.
   Step 4: Choose the best solution.
   Step 5: Build a model or prototype.
   Step 6: Test your solution.
   Step 7: Communicate your solution.
   Step 8: Redesign as needed.
3. a. Step 2: Research
   b. Step 5: Build
   c. Step 7: Communicate
   d. Step 1: Define
   e. Step 4: Choose
   f. Step 8: Redesign
   g. Step 6: Test
   h. Step 3: Brainstorm

DESIGN CHALLENGE 1: MALARIA MELTDOWN!
2. RESEARCH THE PROBLEM
RESEARCH PHASE 1: ANALYZE THE CURRENT MEDICINE CARRIER

1. Temperature is increasing as time goes on.
2. 5.2°/10 minutes = 0.52 degrees per minute; this is the average rate of increase in temperature during the first 10 minutes.
3. 0.3°/10 minutes = 0.03 degrees per minute; this is the average rate of increase in temperature during the last 10 minutes.
4. As time goes on, the temperature increases at a slower rate. In other words, the rate of change in temperature decreases as time goes on.
5. 37°C; the temperature will rise until it matches the temperature of the lab.
6. $15°C \leq T \leq 30°C$

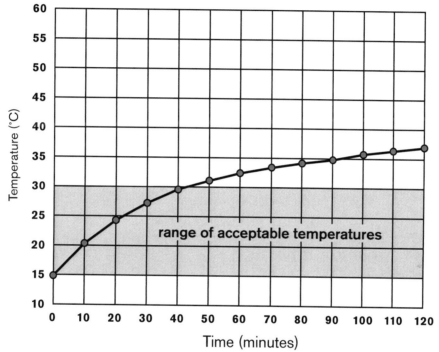

7. In the current medicine carrier, the medicine stays in the acceptable range for just over 40 minutes, at which point the medicine would spoil. This carrier would not be effective enough to bring the medicine all the way to the Yanomami village without it spoiling.
8. Answers will vary. The graph is correct as long as the line remains within the acceptable range. Possible answers:

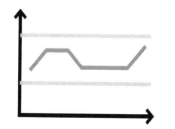 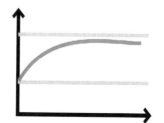 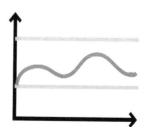

Carrier designs will vary.

RESEARCH PHASE 2: INVESTIGATING DIFFERENT MATERIALS

1. *x*-axis: Time (seconds); *y*-axis: Temperature (°C). Students should connect data points for each material because it is continuous data.

2–4. Answers will vary.

RESEARCH PHASE 3: COMBINING DIFFERENT MATERIALS

1. There are many possible permutations. For materials A and B, permutations include AB, BA, ABA, BAB, AAB, BBA, BAA, ABB, ABAB, AABB, and so forth.

2. Most likely, students will predict that the thickest permutation with the most layers, and the permutation that has more of the material that performed best in Research Phase 2, will be the best insulator.

3–5. Answers will vary.

4. CHOOSE THE BEST SOLUTION

For each material,
(number of layers) • (surface area) • (cost per m²) = total cost

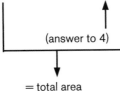

(answer to 4)

= total area

7. COMMUNICATE YOUR SOLUTION

1. Answers will vary.

2. Answers will probably relate to the engineering criteria that the design met. Encourage students to go beyond this to what makes their carrier unique—for example, a handle that makes it easy to carry.

3. Answers will vary. Possible answers: too expensive, too thick, not water-resistant, difficult to carry, and so forth.

4–5. Answers will vary.

DESIGN CHALLENGE 2: MERCURY RISING!
2. RESEARCH THE PROBLEM
RESEARCH PHASE 1: MINIMIZING COST

1. Answers will vary.
2.

DIAMETER OF MERCATRON (CM)	0.4	0.8	1.2	1.6	2.0	2.4
RADIUS OF MERCATRON (CM)	0.2	0.4	0.6	0.8	1.0	1.2
SURFACE AREA OF MERCATRON (CM^2)	0.5024 ≈ 0.5	2.0096 ≈ 2.01	4.5216 ≈ 4.52	8.0384 ≈ 8.04	12.56 ≈ 12.6	18.0864 ≈ 18.09
TOTAL NUMBER OF MERCATRONS NEEDED FOR A TOTAL SURFACE AREA OF 400 CM^2	800	199	89	50	32	23
TOTAL NUMBER OF PACKAGES NEEDED (10 MERCATRONS PER PACKAGE)	80	20	9	5	4	3
COST PER PACKAGE OF 10 MERCATRONS	$0.15	$0.30	$0.60	$1.20	$2.40	$4.80
TOTAL COST	$12	$6	$5.40	$6	$9.60	$14.40

Round answer to next integer above value because you must buy whole Mercatrons and whole packages.

3. 1.2 cm; answers will vary.

RESEARCH PHASE 2: MINIMUM AND MAXIMUM FILTRATION RATES

1. $\dfrac{540 \text{ liters}}{1 \text{ day}} \cdot \dfrac{1 \text{ day}}{24 \text{ hr}} = \dfrac{540 \text{ liters}}{24 \text{ hr}} \cdot \dfrac{1 \text{ hr}}{3600 \text{ sec}} = \dfrac{540 \text{ liters}}{86{,}400 \text{ sec}} \cdot \dfrac{1000 \text{ mL}}{1 \text{ liter}} = \dfrac{540{,}000 \text{ mL}}{86{,}400 \text{ sec}}$

$\dfrac{86{,}400 \text{ sec}}{540{,}000 \text{ mL}} = \dfrac{0.16 \text{ sec}}{1 \text{ mL}} = \dfrac{40 \text{ sec}}{250 \text{ mL}}$

2. $\dfrac{1 \text{ liter}}{1 \text{ min}} \cdot \dfrac{1 \text{ min}}{60 \text{ sec}} = \dfrac{1 \text{ liter}}{60 \text{ sec}} \cdot \dfrac{1000 \text{ mL}}{1 \text{ liter}} = \dfrac{1000 \text{ mL}}{60 \text{ sec}}$

$\dfrac{60 \text{ sec}}{1000 \text{ mL}} = \dfrac{0.06 \text{ sec}}{1 \text{ mL}} = \dfrac{15 \text{ sec}}{250 \text{ mL}}$

3. 15 sec ≤ (time to filter 250 mL) ≤ 40 sec

RESEARCH PHASE 3: INVESTIGATING FLOW RATE

1. Graphs will vary, but should reflect data given in Table 2.3.
2. As diameter increases, flow rate increases (the time it takes to filter 250 mL of water decreases).
3. Answers will vary.
4. Students must interpolate, based on their graphs, the diameter value when time equals 27.5 seconds.
5. Answers will vary. Possible answers: the position of the outlet (how far from bottom); the angle at which water flows out; the cross-sectional area of the outlet spout (if spout rather than hole); the number of outlets.
6. Encourage students to "control" the factor they are testing. If they are testing the number of outlets, they should keep everything the same for each trial and just increase the outlet number by 1 each time (all outlets should be the same size).

7–9. Answers will vary.

PAGE 114

DESIGN CHALLENGE 3: OUTBREAK!
2. RESEARCH THE PROBLEM
RESEARCH PHASE 1: HOW DOES A VIRUS SPREAD?

1. Answers will vary.
2. Everyone in class is typically sick after 5 time intervals.

TIME (DAYS)	0	1	2	3	4	5	6	7	8	9	10	11
NUMBER OF SICK PEOPLE	1	2	4	8	16	32	64	128	256	512	1,024	2,048

3. At each time interval, there are two times as many people sick as in the previous time interval. The numbers of sick people are powers of 2.
4. Let y = number of sick people
   t = time
   $y = 2^t$
5. 1,048,576, or 2^{20}

RESEARCH PHASE 2: THERE'S A DOCTOR IN THE HOUSE

1. Answers will vary.
2.

TIME (DAYS)	0	1	2	3	4	5	6	7	8	9	10	11
NUMBER OF SICK PEOPLE	1	2	3	5	9	17	33	65	129	257	513	1,025

3. At consecutive time intervals, there is 1 less than 2 times the number of sick people. Also, when you compare Table 3.2 to Table 3.1, the number of sick people in Table 3.2 is 1 more than the number of sick people for one earlier time interval in Table 3.1.
4. Let y = number of sick people
   t = time (days)
   $y_{(t+1)} = y_t \times 2 - 1$ or $y = 2^{(t-1)} + 1$ (when $t \geq 1$)

   The first formula is a next-current or recursive formula. It means that to get the next y, represented by $y_{(t+1)}$, you take the current y, represented by y_t, multiply it by 2, and take away 1. This means that if you want to find the number of sick people at time = 20, you have to know all the values that come before 20. The second formula is a direct formula. Given the time (or day number), you can find the number of sick people for that day directly without having to know all the values that come before.
5. 524,289, or $2^{(20-1)} + 1$

RESEARCH PHASE 3: EVERYONE HAS AN AIR-FILTRATION MASK

1. Answers will vary.
2. Data will vary. The number of sick people is typically smaller than in the first two experiments, but the numbers will vary depending on the flip of the coin.
3. The virus should spread more slowly, but eventually everyone will still become sick.
4. In most cases, the air-filtration masks will slow the spread of the virus more than a single doctor.

RESEARCH PHASE 5: COMBINING INTERVENTIONS

1. $0.50 \cdot 0.25 = 0.125 = 12.5\%$
2. $0.50 \cdot 0.04 = 0.02 = 2\%$
3. $0.25 \cdot 0.04 = 0.01 = 1\%$
4. $0.50 \cdot 0.25 \cdot 0.04 = 0.005 = 0.5\%$

6. TEST YOUR SOLUTION

	NO INTERVENTION	MASK ONLY	HAND GEL ONLY	MASK AND HAND GEL
Unvaccinated infection rate	100	50	25	12.5
Vaccinated infection rate	4	2	1	0.5

STUDENT WORK SAMPLE 1

1.   2
2.   Some effort is made to draw in 3-D. The materials are labeled, but drawing is missing dimensions and a net.
3.   Include dimensions, a net, and appropriate units.

STUDENT WORK SAMPLE 2

1.   4
2.   Drawing shows a 3-D view of the design, a net, and additional views of the carrier (a view from the top and a cross section of layers). Materials are labeled, rationale for design is indicated, and all dimensions are labeled—both on the 3-D drawing and the net. Units are appropriate.
3.   Explain why there is a cube that is labeled with 2-cm sides.

STUDENT WORK SAMPLE 3

1.   1
2.   Drawing shows a 3-D view of a prism and the correct corresponding net but doesn't show anything else such as dimensions, materials, units, and so forth. There is no evidence the design criteria were addressed.
3.   Include some mention or labels of materials and dimensions with proper units.

STUDENT WORK SAMPLE 4

1.   3
2.   There is a drawing of the design, but it's not in 3-D. There's a key for showing which material is which, but it's unclear how many layers of each are in the design. The net shows labels of dimensions with units and even shows the calculation of the surface area.
3.   It's unclear why there are two numbers both labeled "surface area." Also, the number of layers of each material is ambiguous in the left-hand drawing. Although the net makes it clear that the carrier is a rectanguar prism, it would be helpful for the left-hand drawing to show a 3-D perspective. Also, the drawing should have labels for the dimensions.